RADICAL ECOLOGY

THE SEARCH FOR A LIVABLE WORLD

CAROLYN MERCHANT

ROUTLEDGE

NEW YORK
LONDON

First published in 1992 by

Routledge
an imprint of
Routledge, Chapman & Hall, Inc.
29 West 35 Street
New York, NY 10001

Published in Great Britain by

Routledge
11 New Fetter Lane
London EC4P 4EE

Acknowledgment is made for the permission of *Tikkun Magazine* to include portions of the author's article, "Gaia's Last Gasp," (March, April 1990); of HarperCollins Publishers to include portions of the author's book *The Death of Nature: Women, Ecology, and the Scientific Revolution* (1980) and portions of "Preface 1990," (second edition, 1990); of *Environmental Ethics* to include portions of the author's article "Environmental Ethics and Political Conflict," (Spring 1990); of The University of Wisconsin Press to include portions of the author's article "Restoration and Reunion with Nature," *Restoration and Management Notes* (Winter 1986); of the University of North Carolina Press to include the caption and diagram "Theoretical Framework for Interpreting Ecological Revolutions," from *Ecological Revolutions: Nature, Gender, and Science* (1989), and to Sierra Club Books to include portions of the author's article "Ecofeminism and Feminist Theory," and accompanying diagram from *Reweaving the World: The Emergence of Ecofeminism*, ed. Irene Diamond and Gloria Orenstein, 1990.

Library of Congress Cataloging in Publication Data

Merchant, Carolyn.
    Radical ecology : the search for a livable world / Carolyn
Merchant.
        p.   cm.—(Revolutionary thought/radical movements)
    Includes bibliographical references (p.   ) and index.
    ISBN 0-415-90649-0 (cloth)   ISBN 0-415-90650-4 (paper)
    1. Ecology—Philosophy. 2. Ecology—Political aspects. 3. Human
ecology—Moral and ethical aspects. 4. Ecofeminism.   I. Title.
II. Series.
QH540.5.M48   1992
304.2—dc20   92-12542
                                                        CIP

British Library Cataloguing in Publication Data also available.

RADICAL
ECOLOGY

Revolutionary Thought/ Radical Movements

A book series edited by Roger S. Gottlieb

Other books in the series:

The Gay and Lesbian Liberation Movement
Margaret Cruikshank

Marxism 1844–1990: Origins, Betrayal, Rebirth
Roger S. Gottlieb

Women in Movement: Feminism and Social Action
Sheila Rowbotham

To the Earth

CONTENTS

SERIES EDITOR'S PREFACE   xi

ACKNOWLEDGMENTS   xvii

INTRODUCTION: WHAT IS RADICAL ECOLOGY?   1

     Self in Society   2

     Society in Self   4

     Self Versus Society   7

     Radical Ecology   9

I   PROBLEMS

1.   THE GLOBAL ECOLOGICAL CRISIS   17

     Air; Water; Soils; Biota   18

     Political Economy   23

     Environmental Problems in the Second World   26

     Population   29

     Steady-State Economics   37

     Conclusion   38

     Further Reading   39

2   SCIENCE AND WORLDVIEWS   **41**

The Organic Worldview   42
The Rise of Capitalism   44
Experimental Science   45
The Mechanistic Worldview   48
The Domination of Nature   54
Newtonian Science   55
Conclusion   58
Further Reading   59

3. ENVIRONMENTAL ETHICS AND   **61**
POLITICAL CONFLICT

Egocentric Ethics   63
Homocentric Ethics   70
Ecocentric Ethics   74
Conclusion   81
Further Reading   81

II   THOUGHT

4. DEEP ECOLOGY   **85**

Principles of Deep Ecology   86
Scientific Roots of Deep Ecology   93
Eastern Philosophy   100
Critiques of Deep Ecology   102
Reconstructive Science   105
Conclusion   107
Further Reading   108

5. SPIRITUAL ECOLOGY   **110**

The Council of All Beings   111
Nature Spirituality   113
The Old Religion   117
Native American Land Wisdom   120
Mainstream Religions   122
Ecological Creation Spirituality   124
Ecological Process Theology   126

Conclusion   129
Further Reading   129

6.   SOCIAL ECOLOGY   132

Progressive Ecology: Marx Meets Muir   133
Marx and Engels on Ecology   134
Anarchist Social Ecology   142
Socialist Ecology   146
Dialectical Biology   150
Conclusion   153
Further Reading   154

III   MOVEMENTS

7.   GREEN POLITICS   157

The Group of Ten   159
Populism   162
Minority Activists   164
The Greens   167
Greens in the Second World   171
Earth First!   173
Greenpeace   176
Direct Action   177
Conclusion   180
Further Reading   181

8.   ECOFEMINISM   183

Liberal Ecofeminism   188
Cultural Ecofeminism   190
Social Ecofeminism   194
Socialist Ecofeminism   195
Socialist Ecofeminism and Production   197
Socialist Ecofeminism and Reproduction   198
Women in the Third World   200
Women in the Second World   207
Conclusion   209
Further Reading   209

9. SUSTAINABLE DEVELOPMENT 211

Sustainable Agriculture 212
Biological Control 215
Restoration Ecology 216
Bioregionalism 217
Indigenous Peoples and Sustainability 222
Global Sustainable Development 227
Conclusion 232
Further Reading 233

CONCLUSION: THE RADICAL ECOLOGY 235
MOVEMENT

Contributions of Radical Theorists 236
Contributions of Radical Activists 236

NOTES 241

INDEX 263

SERIES EDITOR'S PREFACE

This book, like its companions in the *Revolutionary Thought/Radical Movements* series, challenges contemporary society and civilization.

Perhaps the heart of this challenge is a deeply felt anguish and outrage over the sheer magnitude of human suffering—along with the terrible frustration of knowing that much of this suffering could be avoided. Radicals refuse to blame homelessness and starvation, the rape of women and abuse of children, the theft of labor and land, hope and self-respect on divine Providence or unchangeable human nature. Rather, they believe that much of it comes from injustice, exploitation, violence, and organized cruelty that can be eradicated. If we drastically alter our social arrangements in the direction of equality, justice, and human fulfillment, the brutal realities of the present can give way to vastly increased material security, social harmony, and self-realization.

Philanthropists and political reformers share radicals' concern for human suffering. But unlike reformers and philanthropists, radicals and revolutionaries address whole *systems* of injustice. In these systems, particular groups are humiliated, denied rights, subject to unjust control. The few become rich while the many suffer from poverty or economic insecurity. The select get privileges while millions learn submission or humiliation. We are conditioned to false needs for endless consumption while nature is poisoned. The powers-that-be profit from these systems, "common sense" enshrines them as necessary, and ideological mystification obscures their origin and nature by blaming the victims. Responses to people's pain, if they are to be truly and lastingly effective, must be aimed *at the system:* at capitalism, sexism, racism, imperialism, homophobia, the bureaucratic state, and the domination of nature.

Governments and economies, families and culture, science and individual psychology—all are shaped by these systems of domination and exclusion. That is why the radical ideal goes beyond piecemeal improvements to a Utopian vision; and tries to realize that vision in everyday struggles for a fair distribution of power, human dignity, and a livable environment. Revolutionaries have argued that a modern economy can be democratically controlled and oriented to human needs rather than profit; can do without vast differences of wealth and power; and can preserve rather than destroy the earth. Radicals claim that in a true 'democracy' ordinary men and women would help shape the basic conditions which affect their lives: not just by an occasional trip to the ballot box, but by active involvement in decisions about political and economic life.

How will these sweeping changes take place? Revolutionaries have offered many answers—from large political parties to angry uprisings, from decentralized groups based in consciousness-raising to international organizations. In any case, however, the conception of radicalism which informs the series stipulates that authentic revolutionary change requires the self-action of sizable groups of people, not the self-promotion of a self-proclaimed revolutionary "elite." The only way to prevent the betrayal of the revolution by a privileged bureaucracy is to base radical politics on free discussion, mutual respect, and collective empowerment *from the beginning.* This is one of the clearest and most painful lessons from the history of communism.

Of course much of this sounds good on paper. Yet it may be—as many have claimed—that radical visions are really unrealistic fantasies. However, if we abandon these visions we also abandon human life to its current misery, with little to hope for but token reforms. Radicals reject this essentially cynical "realism," opting for a continuing faith in the human capacity for a fundamentally different and profoundly liberating form of life.

In fact, people have always dreamed of a better world. Yet it is only since the late eighteenth century that organized groups developed a systematic theoretical critique of social life; and tried to embody that critique in mass political movements designed to overthrow the existing order of economic ownership and political control. American revolutionaries claimed that "All men are endowed with certain inalienable rights." The French revolution demanded "liberty, equality, fraternity."

Since then Marxist, socialist, feminist, national liberation, civil rights, gay and lesbian liberation, and ecology movements have been born. Each movement utilized some of the accomplishments of its predecessors, criticized the past for its limitations, and broke new ground. *Revolutionary Thought/Radical Movements* will focus on the theory and practice, successes and failures, of these movements.

While the series' authors are part of the radical tradition, we are painfully aware that this tradition has committed grave errors and at times failed completely. The communism of the Eastern bloc, while maintaining certain valuable social welfare programs, combined economic inefficiency, brutal tyranny, and ecological devastation. Many of us who took to the streets in the 1960s joined arrogance with idealism, self-indulgence with utopian hopes. Much of contemporary radical or socialist feminism fails to reach beyond a circle of the already converted.

These and other failures of radicalism are certainly apparent today. Daily headlines trumpet the collapse of the Eastern bloc, the US victory in the Cold War, the eternal superiority of capitalism and free markets, and the transformation of yesterday's radicals into today's yuppies. Governments of countries that had called themselves "socialist" or "communist" (however much they were distorting the meaning of these terms) trip over each other rushing west for foreign corporate investment and economic advice.

But there are also *successes,* ways in which radicals have changed social life for the better. Though these achievements have been partial reforms rather than sweeping revolutions, many of the basic freedoms, rights, and material advantages of modern life were fought for by people called radicals, dangerous revolutionaries, or anti-American:

- restrictions on the exploitation of workers, from the eight-hour day to the right to unionize;
- resistance to cultural imperialism and racial discrimination;
- a host of government programs, from unemployment insurance to social security, from the Environmental Protection Agency to fair housing laws;
- restrictions on opportunistic and destructive American foreign policy in Vietnam, El Salvador, Nicaragua, and other nations.

While radicals have not been alone in seeking these goals, they have often led the fight. Perhaps more important, they have offered a theoretical analysis which shows the *connections* between problems which may appear to be separate. They have argued that the sexist treatment of women and ecological devastation may have the same root. They have shown the links between the private control of wealth and an expansionist foreign policy. They have analyzed the family, the factory, the army, and the government as parts of the same system of domination.

Along with both the concrete successes and the global vision, radicals have—sadly—too often reproduced the ideas and relationships they sought to destroy. Marxists demanded an end to unjust society—yet formed authoritarian organizations where dissent was repressed. Radical feminists proclaimed "sisterhood is powerful," but often ignored Black women or poor women. At times ecologists, in trying to save nature, have been disrespectful of human beings.

Some of the worst failures came, in short, not from being radical, but from *not being radical enough:* not inclusive enough, not honest enough, not willing to examine how radical political programs and group behavior reproduced an oppressive, unjust society. Awareness of these failures reminds us that revolutionary thought cannot limit itself to critique of the larger society, but also requires self-criticism. While this process can degenerate into petty sectarian hostilities, it also shows that authentic radicalism is not a dead graven image, but a living quest to learn from the past and change the future. In the attempt to create solidarity and community among the oppressed, for instance, radicals have recently spent much effort trying to address and appreciate fundamental differences in social experience—between black and white workers, men and women, temporarily able-bodied and disabled, gay/lesbian and straight. In this effort, radicals have wrestled with the paradox that persons may simultaneously be victims of one system of domination and agents of another one.

The books in this series are part of this radical quest for revolutionary change and continued self-examination. In an era of the sudden fall of totalitarian communism and the frightening rise in the federal deficit, of the possibility of a peace dividend and the specter of the death of nature—these discussions of revolutionary thought and radical movements are needed more than ever before.*

Roger S. Gottlieb

*Thanks for editorial suggestions to Bland Addison, Mario Moussa, Miriam Greenspan, Tom Shannon and John Trimbur.

ACKNOWLEDGMENTS

Many people, as well as other organisms and the entire planet, have made this book possible. Special thanks are expressed to Roger Gottlieb and Lisa Freeman who helped to conceptualize the book and its contents. My colleagues in Conservation and Resource Studies at the University of California, Berkeley, friends in the environmental community here and around the world, and students in my Environmental Philosophy and Ethics classes have stimulated, supported, and constructively criticized many of the ideas in it. I am especially grateful to Yaakov Garb, Florence Gardner, and Tamara Whited for research assistance and Celeste Newbrough who prepared the index. Responsibility for the final form of the ideas and arguments is of course my own.

C. M.
Berkeley, California
January 1992

INTRODUCTION:
WHAT IS RADICAL ECOLOGY?

Radical ecology emerges from a sense of crisis in the industrialized world. It acts on a new perception that the domination of nature entails the domination of human beings along lines of race, class, and gender. Radical ecology confronts the illusion that people are free to exploit nature and to move in society at the expense of others, with a new consciousness of our responsibilities to the rest of nature and to other humans. It seeks a new ethic of the nurture of nature and the nurture of people. It empowers people to make changes in the world consistent with a new social vision and a new ethic.

To become clear about our own goals for change, we need to reflect on the ways in which we have absorbed the norms and roles of the larger society in which we live. How can we replace feelings of individual helplessness with feelings of power to make changes consistent with a new social vision and a deeper, more articulate environmental ethic? We can begin by reflecting on our own family history and our own socialization.[1]

SELF IN SOCIETY

Consider your own family's history and place in society going back at least to your grandparents' generation. Were your ancestors native to this country? Are you or your parents first-, second-, or perhaps eighth-generation immigrants? What large events—wars, depressions, revolutions, social movements—shaped their lives? How did your families use the land and relate to nature? Which of their values have you absorbed? Which have you rejected? Think also about the people you know and their family connections to the land.

As people ponder these questions, they become aware of deep-seated contradictions in the ways different classes of people use the land, the ways in which their own values are shaped by their family's history, and of their own struggles to develop new ways of interacting with nature. One student in an environmental ethics class writes of her emerging consciousness about the land as source of both commodity and beauty, of people as both beneficiaries of and laborers on the land, and the work ethic that has guided individuals in the struggle to overcome hardship.

> I grew up with my father's extended family. His family is mostly made up of farmers. My father grows wine grapes. His uncles are rice farmers; his Aunt Opal is an Oklahoman who came out to California during the dust bowl years. My father is "the one who made it" on his side of the family. I grew up pruning grapes alongside my uncle and Mexican migrant workers on weekends and attending good schools with affluent Marin County kids on weekdays. I spent many years working with my brothers and sisters and family friends out in the fields, picking grapes, pruning, installing irrigation systems, suckering, tying vines, or rounding up cattle and sheep. While working in the fields I grew to respect and wonder at nature. . . . Since my background is ethnically diverse, I was raised without specific religious or ethnic indoctrination. My life experience has created values oriented around family, hard work, interaction with nature, education, and contribution to society.

Another student's family history seems to recapitulate American history in optimizing opportunities presented by the westward movement in a land of abundance. Yet this same expansionist potential apparent to her nineteenth century ancestors poses a sharp contradiction for her own twentieth century consciousness shaped by a growing sense of the need for conservation and an alternative land ethic.

My mother's family descended from French Huguenots who fled to England and then came to the United States around the time of the American Revolution. Their Puritan work ethic and pioneer spirit, searching for abundant land resources and freedom, are the foundation upon which the values of my mother's family developed. Their family settled in the Tennessee hills. Later my great-great-grandfather made the move by train and covered wagon to the promised land of the Oregon country. My mother's side of the family were farmers and its seems that each successive generation worked its way up the socioeconomic ladder through hard work and thrift and an especially strong emphasis on education. Both my grandmother and grandfather obtained college educations which were made possible by the land grant and agricultural college systems. My grandfather's farming practices were influenced by the new conservation practices emerging in the 1950s in reaction to the devastating effects of the Oklahoma Dustbowl.

The ethics I struggle with today have evolved within the context of a family history whose relationship has been close to the earth. I believe that today we need a more spiritual way of feeling our relationship with the earth. I also believe an environmental ethic must acknowledge the historical domination of women and the environment by men. Our ethical model needs to come from outside patriarchal social structures.

A recent student immigrant reflects on his family's class status, recognizing the ways in which privilege in both First and Third World countries is linked to exploitation of the earth and other humans. He intimates that radical transformation is needed to reverse the failures of social justice and the degradation of the planet.

Born in Lima, into the richest ruling class of Peru, my family of eight (six kids, I the baby) was capable of escaping the political persecution of the incoming military regime in 1969 by moving to an entirely new area in California—the land of suburbia. There our familial Catholic, South American, upper class morals and behavior would be coupled with the surrounding WASP, upper-middle class, consumptionist mentality to create my socio-economic environment. . . . Wastefulness, materialism, and inequality were accepted and ubiquitous, while conservation, non-material wealth and happiness, and social justice were inheard of and unpursued. . . . The notion that the planet earth had seen better days, which dawned upon me gradually in high school, gained momentum in college. About the time my sister joined Earth First!, I took an environmental studies class and all my worst fears were confirmed.

All three students have become painfully aware that the transformation of nature into commodity, which allowed their parents' and grandparents' generations to rise in status, has had immense linked environmental and human costs. The value placed on the individual's hard work brought family success, but new values that sustain rather than degrade nature and other people are now needed.

SOCIETY IN SELF

How have you yourself been socialized? What effect has the society in which you grew up had on you as a female or male? Have you experienced sexism or racism in your daily life? What historical forces—immigration movements, urbanization, social mobility, educational opportunities—have helped to create your own economic position? Think about the values you have derived from your school, your church, and your workplace. How have the politics and economics of your community affected you? What environmental values have you formed as a result?

One student contrasts his family's economic socialization in rural and urban China with his own socialization in New York City. Immigration, he believes, fostered frugality and conservation of economic resources in the city, a value derived from peasant life in rural China. He ponders whether he can transcend his urban socialization to reclaim the connections his grandmother once felt to nature.

My grandparents spent most or all of their adult lives as peasant farmers in small villages outside of Canton, China. My parents moved out of rural China during the 1950s and 1960s to come to New York City. My father managed a Cuban-Chinese restaurant while my mother occasionally took in sewing. From my grandparents to my parents to me, my family has moved through dimensions of geography, nationality, culture, industrialization, and financial power—from rice paddies in rural China to the industrialized islands of Hong Kong and Cuba to the richly dense urban construction and development of New York City.

What environmental ethics and philosophy have arisen from such a background? There is an ethic of non-wastefulness; to take only as needed, to conserve as much as possible, to put to use as many aspects of a resource as possible. There is a sense that resources are limited—there is only so much water in the world, only so much money in the family. Perhaps these two ethics come from the habits of subsistence living, the life of a peasant farmer, the life of lower-class immigrants just arrived in a new land. The little resources not required for subsistence are saved up, for two future possibilities: the opportunity to rise up from subsistence living, and the possibility of disaster, of a disruption in the flow of resources. In the rice paddies, there were no pesticides, no processed fertilizers, no weather satellites—you used your senses and your body and your memory. I've sometimes envied my maternal grandmother, because it seems as if she is in some special harmony with the world—a harmony which I feel is lost to me, a member of a very industrialized society whose experience of this world is heavily mediated by technology. Yet I do feel that my maternal grandmother connects me with that harmony. But how will I, in turn, pass it on to my own children, should I have any?

Another student is deeply aware of how her own place in society has been carved out for her through a long past history of male interests and influence in the economy and politics. She reflects on how both men and women are shaped by society's expectations of them, creating roles that incorporate the dominant worldview in which humans are individual atoms in a vast social and cosmic machine. She suggests that only a painful inner transformation to a new ethic will allow people to move beyond historically created roles to realize an ecologically just society.

I am the granddaughter of four European immigrants. My parents' highest value is upward mobility. Their personal history, as the only son of urban Italians and as a girl longing to get off the farm on the edge of the Dustbowl during the Great Depression, sealed their "ethical fate." In my family, my father's ethnic heritage is dominant. Boys are preferred over girls. Patriarchal values and a pervasive sense of guilt have dribbled down. Individual achievement is paramount. If you are a girl, you must either choose the female traditional role or your achievement must be of an even higher order than boys. In a broader sense, my family's philosophies are mechanistic. We are only cogs in a great machine—the individual soul and personality have no intrinsic worth. Since this implies that we are all interchangeable, we must be competitive in *every* situation, lest another take our place. Self-interest is the highest priority. My familial experience has propelled me to the left. For me ecojustice seems to be the truth. As Robertson Davies said (in the *Deptford Trilogy*), "If you do not choose a philosophy of life (however painful that choice may be), it will choose you."

Through reflection, another woman becomes aware that her socialization is the outcome of a combination of the economic forces shaping a company town that exploited men as miners and nature as a resource and an unusual "feminized" Methodist religious heritage. Her environmental ethic is a consequence of the freedom of thought this feminist heritage fostered.

My great-grandparents on my father's side of our family lived in a coal mining town outside of Morgantown, West Virginia. It was a 'company-town,' owned and controlled by one man. Great-grandfather was one of the miners and he and my great-grandmother lived in what my Dad has referred to as a 'shack.' After my great-grandfather died of black lung, my grandfather began working in the mines. But the mining town split up around the 1920s. I guess that they had extracted all of the existing coal from that area of the Blue Ridge Mountains.

My parents had no sons, so their four daughters played the roles of girl and boy, daughter and son. We were all raised in the Methodist Church, attending Sunday School as children and Methodist Youth Fellowship as teenagers. Our church had three women pastors. Lay women would often read scriptures during services, and references to "he" in the Bible were always read "he/she." I was always encouraged to formulate my own religious ideas and eventually rejected Christianity altogether. . . . My environmental ethic began with a gut level reaction to environmental destruction, supporting its wrongness with facts, and developing a new set of morals and values to live by.

These voices reveal some of the ways in which social patterns are imprinted on us as we grow up amid a variety of economic, political, religious, and genderized social forces. Recognizing that we ourselves are reflections of the values and norms of the larger society allows us to step back and reassess those values. Through this process we can articulate an ethic that either sustains or reforms the institutions around us. In so doing, however, we may find ourselves acting at odds with the dominant values of our society.

SELF VERSUS SOCIETY

Our lives today bear the continuities of the past, but our futures reflect the problems facing the next generations. We go on making and remaking ourselves each day as history unfolds and society changes. What conflicts do you experience between your own values and goals and the institutions and environment you anticipate in the future? What expectations do you have for yourselves and your children? How might your children's values differ from your own? How can you help to bring about a world that will provide them with a high quality of life?

One student professes skepticism that the underlying capitalist system can be transformed, but offers education as a method of revealing its inherent contradictions and a pathway toward reforming its problems.

Until I went to boarding school my world was very simple. There were the bad people who strip-mined for coal and there were the good people who ran my summer camp. School forced me to question my basic assumptions. Suddenly my black and white world was overwhelmingly grey. It did not occur to me that minimum impact camping was possible only if you accepted the fact that a Third World country was being drilled and drained of its precious blood. The contradictions in society are everywhere. It is too easy to reduce the present world situation to good guys and bad guys. I recognize that capitalism has problems, but it does not seem productive to label the system as the scapegoat for all of society's ills. I look to education to remedy the problems facing the world, because I am convinced that if people understand what is happening they will work toward a solution.

Another student places hope in social movements as a transformative method. She sees her own alienation from society as a source of power that enables her to find others willing to work toward meaningful change.

Once I believed I was beyond the influence of class structure, a hybrid cross with the ability to choose my class identity. But life has a way of obliterating fantasy. Currently, I have no difficulty identifying myself as working class, although the average person might see me a part of the great American middle-class. My first awareness of myself as opposed to society was the early knowledge of my bisexuality and I viewed society and its dominant institutions through an outcast's eyes. I now understand myself within the context of alienation, of self versus society. However, I have mitigated this stance by initiating and participating in group actions to change the institutions I find alienating. Through participation in movements, I have been able to experience, even to create, society in self, while acting as self against society.

Is there a way to move forward both in thought and action that diminishes feelings of helplessness as well as tendencies to "blame the system?" Can we find a ground for environmental analysis and a means for putting it into practice? Radical ecology offers one such approach— an approach that helps us to analyze current problems, to construct new theoretical frameworks, and to find people and social movements that support our efforts to improve the quality of life.

RADICAL ECOLOGY

Ecology as a science emerged in the late nineteenth century in Europe and America, although its roots may be found in many other places, times, and cultures. The science of ecology looks at nonhuman nature, studying the numerous, complex interactions among its abiotic components (air, water, soils, atoms, and molecules) and its biotic components (plants, animals, bacteria, and fungi). Human ecology adds the interactions between people and their environments, enormously increasing the complexities. Human ecology has been most successful when it studies clearly defined places and cultures—the Tsembaga people of Papua New Guinea, the Shoshone Indians of the American west, the Tukano Indians of the Amazonian rainforests. When time is added as an additional dimension, environmental history emerges as a subject. Even here, temporal changes in specific regions have provided the most grist for the mills of environmental historians—the ecological history of New England, the emergence of hydraulic society in California, changing ideas of wilderness and conservation in America, and so on.

Social ecology takes another step. It analyzes the various political and social institutions that people use in relationship to nature and its resources. Technologies—such as axes, guns, and bulldozers—transform trees, animals, and rocks into "natural resources." Systems of economic production, such as hunting, gathering, and fishing, subsistence agriculture, and industrial manufacturing turn the resources into goods for home use or market trading. Cultural systems of reproduction provide norms and techniques that guide families in deciding whether and when to bear children. Laws and politics help to maintain and reproduce the social order. Ideas and ideologies, such as myths, cosmologies, religion, art, and science, offer frameworks of consciousness for interpretating life and making ethical decisions.

Radical ecology is the cutting edge of social ecology. It pushes social and ecological systems toward new patterns of production, reproduction, and consciousness that will improve the quality of human life and the natural environment. It challenges those aspects of the political and economic order that prevent the fulfillment of basic human needs. It offers theories that explain the social causes of environmental problems and alternative ways to resolve them. It supports social movements for removing the causes of environmental deterioration and raising the quality of life for people of every race, class, and sex.

How can radical ecology help to bring about a more livable world? Environmental problems, as I argue in Part I, result from contradictions (tendencies to be contrary to each other's continuance) in today's society. The first contradiction arises from tensions between the economic forces of production and local ecological conditions, the second from tensions between reproduction and production: The particular form of production in modern society—industrial production, both capitalist and state socialist—creates accumulating ecological stresses on air, water, soil, and biota (including human beings) and on society's ability to maintain and reproduce itself over time.

The first contradiction arises from the assaults of production on ecology. Examples include the destruction of the environment from the uses of military production (such as the oil spills and air pollution during the 1991 Gulf War or the predicted nuclear winter from nuclear war); global warming from industrial emissions of carbon dioxide; acid rain from industrial emissions of sulphur dioxide; ozone depletion from industrial uses of chlorofluorocarbons; the pollution of oceans and soils from the dumping of industrial wastes; and industrial extractions from forests and oceans for commodity production. These assaults of production on global ecology are circulated by means of the biogeochemical cycles and thermodynamic energy exchanges though soils, plants, animals, and bacteria (see Figure I.1, center circle). Their effects are experienced differently in the First, Second and Third Worlds and by people of different races, classes, and sexes.

The second contradiction arises from the assaults of production on biological and social reproduction. The biological (intergenerational) reproduction of both human and nonhuman species is threatened by radiation from nuclear accidents (such as the 1979 accident at Three Mile Island in the United States and the 1986 accident at Chernobyl in the Soviet Union) and by toxic chemicals from industrial wastes. The reproduction of human life on a daily (intragenerational) basis in Third World countries is endangered as local food, water, and fuel supplies are depleted by the conversion of lands to cash crops and in the First World as harmful chemicals in foods, drinking water, and indoor air invade the home. The reproduction of society as a whole is imperiled by government policies that support continued industrial pollution and depletion and by industry policies that support continued sex and race discrimination (see Figure I.1, middle circle). A country's form of social reproduction and its form of economic production constitute its political economy. Thus the United States, China, Brazil, Kenya, and Malaysia all have particular political economies.

The global ecological crisis of the late twentieth century, I argue, is a result of these deepening contradictions generated by the dynamics between production and ecology and by those between reproduction and production. But problems of pollution, depletion, and population expansion have specific roots in each country's internal history, its place in the global order, and the current trajectory of its internal development. Each environmental problem therefore needs to be examined in the context of its own specific history as well as its linkages to global political economies (Chapter 1).

As these two contradictions become more visible, they also undermine the efficacy of western culture's legitimating worldview, pushing philosophers, scientists, and spiritualists to rethink human relationships with the nonhuman world (see Figure I.1, outer circle). The mechanistic worldview created during the seventeenth century scientific revolution constructs the world as a vast machine made up of interchangeable atomic parts manipulable from the outside, just as the parts of industrial machines can be replaced or repaired by human operators. This mechanistic worldview, which arose simultaneously with and in support of early capitalism, replaced the Renaissance worldview of nature as a living organism with a nurturing earth at its center. It entailed an ethic of the control and domination of nature and supplanted the organic world's I–thou ethic of reciprocity between humans and nature. Mechanism and its ethic of domination legitimates the use of nature as commodity, a central tenet of industrial capitalism (Chapters 2 and 3).

Deep ecologists (Chapter 4) call for a total transformation in science and worldviews that will replace the mechanistic framework of domination with an ecological framework of interconnectedness and reciprocity. Spiritual ecologists (Chapter 5) see the need to infuse religions with new ecological ideas and revive older ways of revering nature. Social ecologists (Chapter 6) see a total transformation of political economy as the best approach. Most of these theories entail an ecocentric ethic in which all parts of the ecosystem, including humans, are of equal value, or an ecologically-modified homocentric ethic that values both social justice and social ecology.

Radical environmental movements draw on the ideas and ethics of the theorists, but intervene directly to resolve the contradictions between ecology and production and between production and reproduction. Green political activists (Chapter 7) advocate the formation of green parties that would recast social and political reproduction and a variety of direct actions that would reverse the assaults of production on reproduction by saving other species, preserving human health, and cleaning up the environment. Ecofeminists (Chapter 8) are particularly

FIGURE I.1
ECOLOGICAL REVOLUTIONS

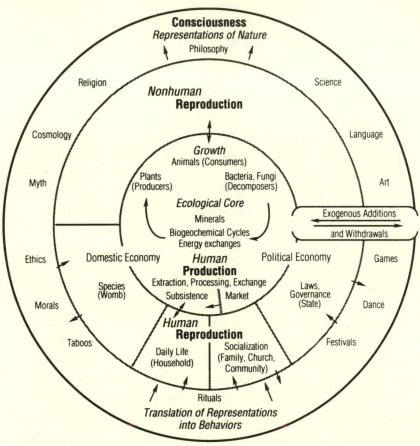

Source: Carolyn Merchant, *Ecological Revolutions: Nature, Gender, and Science in New England*. The University of North Carolina Press, Chapel Hill and London, 1989; pp. 6–7, reprinted by permission.

concerned about issues that affect women's own bodies in biological reproduction (such as toxic substances and nuclear radiation) and women's roles in social reproduction (such as altering workplace/homeplace patterns and norms). The sustainable development movement (Chapter 9) searches for new approaches to resource use that would reverse

Figure I.1 Conceptual Framework for Interpreting Ecological Revolutions. Ecology, production, reproduction, and consciousness interact over time to bring about ecological transformations. The innermost sphere represents the ecological core within the local habitat, the site of interactions between ecology and human production. Plants (producers), animals (consumers), bacteria and fungi (decomposers), and minerals exchange energy among themselves and with human producers in accordance with the laws of thermodynamics and the biogeochemical cycles. Introductions and withdrawals of organisms and resources from outside the local habitat can alter its ecology. Human production (the extraction, processing, and exchange of resources and commodities) is oriented toward immediate use as food, clothing, shelter, and energy for subsistence or toward profit in mercantile trade and industrial capitalism. With increasing industrialization, the subsistence-oriented sector declines and the market-oriented sector expands (as indicated by the clockwise arrow).

The middle sphere represents human and nonhuman reproduction. The intergenerational reproduction of species and intragenerational survival rates influence ecological interactions directly in the case of nonhuman individuals or as mediated by production in the case of humans. In subsistence (or use-value) societies, production is oriented toward the reproduction of daily life in the household through the production of food, clothing, shelter, and energy (as indicated by the two-way arrow). For humans, the reproduction of society also includes socialization (in the family, church, and community) and the establishment of laws and governance that maintain order in the tribe, town, state, or nation.

Human consciousness, symbolized by the outermost sphere, includes representations of nature reflected (as indicated by the arrows) in myth, cosmology, religion, philosophy, science, language, and art, helping to maintain a given society over time and to influence change. Through ethics, morals, taboos, rituals, festivals, the dance, and games, they are translated into actions and behaviors that both affect and are affected by the environment, production, and reproduction (as indicated by the arrows).

The "semipermeable" membranes between the spheres symbolize possible interactions among them. Ecological revolutions are brought about through interactions between production and ecology and between production and reproduction. These changes in turn stimulate and can be simulated by new representations of nature and forms of human consciousness.

the assaults of production on ecology, thereby renewing and preserving soils, waters, air, and biota.

Although radical ecology pushes for change and social transformation, it is not a monolithic movement. It has many schools of thought and many action groups. Its branches are often at odds in goals and values, as well as techniques and specific actions. These produce conflicts and heated debates within the larger movement resulting in a variety of approaches to resolving environmental problems. My own view is one of guarded optimism, placing hope in social movements that intervene at the points of greatest ecological and social stress to reverse ecological damage and fulfill people's basic needs. The goals of production need to be subordinated to the reproduction of life through the fulfillment of human needs and the preservation of local ecologies and be informed by an ethic of partnership between humans and nonhuman nature. Although the new worldview advocated by deep and spiritual ecologists may not lead the social transformation, it can nevertheless foster and support the new economic and social directions taken. Perhaps over the next five decades a global ecological revolution will take place so that by the middle of the twenty-first century we will have new forms of production, reproduction, and consciousness that will sustain both people and the natural environment. Such a transformation would fulfill much of the vision and hope of radical ecology.

Many people will disagree with the goals of radical ecology. Perhaps most will decline to participate in its various actions. Yet radical ecology offers a critical standpoint from which to view and analyze mainstream society and mainstream environmentalism. It sharpens our understanding of the assumptions underlying Western civilization and its values. It broadens our perspective on Second and Third World economic and environmental problems. It helps us to formulate answers to the dilemmas of self in society, society in self, and self versus society.

The visibility of radical environmental movements may make mainstream environmental goals more acceptable. Radical actions often raise public consciousness about issues enmeshed in bureaucratic technicalities. Changes triggered by radical actions may then come about through normal political processes. Although it may fail to bring about revolutionary transformation, radicalism can still be effective in changing attitudes, raising consciousness, and promoting social change. The following chapters offer an account of environmental problems, radical ecological theories, and social movements from the perspectives of both proponents and critics in the search for a livable world.

I

PROBLEMS

I

PROBLEMS

1

THE GLOBAL ECOLOGICAL CRISIS

The world of the late twentieth century is experiencing a global ecological crisis, one that is both a product of past ecological and economic patterns and a challenge for the future. From Chernobyl radiation to the Gulf War oil spill; from tropical rainforest destruction to polar ozone holes; from alar in apples to toxics in water, the earth and all its life are in trouble. Industrial production accentuated by the global reproduction of population, has put stress on nature's capacity for the reproduction of life. Pollution and depletion are systematically interlinked on a scale not previously experienced on the planet.

As we approach the millennium of the twenty-first century, perceptions of planetary destruction and calls for the earth's renewal abound. Can planetary life sustain itself in the face of industrial assaults? How is the current environmental crisis in production manifested? How are the planet's airs, waters, soils, and biota interconnected? How might life be restored to the planet? A new partnership between humans and nonhuman nature is needed.

During the past decade the dimensions of a global ecological crisis have become painfully visible. In January 1989, *Time* magazine's person of the year award went to "The Endangered Earth," graphically illustrated by sculptor Christo as a suffocating globe wrapped in plastic and bound with twine. With increasing public awarness of global problems, public concern has mounted. The Alaskan oil spill alerted millions to the tragic transformation of a pristine Alaskan shoreline surrounded by lush rainforest into black, motionless, silent beaches of dead birds, seals, sea otters, and contaminated waters devoid of sustenence for local fishers and their families. In June 1989, a *New York Times*/CBS poll found that an astonishing 80 percent of all Americans questioned overwhelmingly agreed with the statement: "Protecting the environment is so important that requirements and standards cannot be too high, and continuing environmental improvements must be made regardless of cost."[1]

AIR

Today the hot air of the "greenhouse gases" threatens atmospheric chemistry balances. As the amount of carbon dioxide and other gases in the atmosphere increases from the industrial processes and the burning of fossil fuels, global temperatures are predicted to rise from 3 to 10 degrees Fahrenheit over the next century. Perhaps the most widely-felt evidence of global warming was the intense hot weather experienced by Americans during the summer of 1988. "The greenhouse effect is already here and it will worsen," warned scientists and policy analysts at Congressional hearings held that summer. According to Senator Timothy Wirth, "The greenhouse effect is the most significant economic, political, environmental, and human problem facing the

21st century."[2] Three countries, the United States (21 percent), the USSR (19 percent), and China (10 percent) together produce 50 percent of all carbon dioxide emissions.[3] With the greenhouse effect, winters would become stormier, summers hotter and drier. Seas could rise one to three feet over the next half century; hurricanes would become more powerful as the oceans warm. Waterfront homes will be flooded, midwestern droughts will increase in severity, grain growing regions will move north, and whole forests and wild species will be lost.[4] Although there is much debate over the timing of the effect, a series of measures to slow it have been recommended, such as stopping global deforestation, planting trees, conserving heating fuel, and shifting to alternative energy sources.[5]

Ozone depletion is another global disruption caused by industrial production. In 1985 scientists reported a hole in the ozone layer over the Antarctic. As a result of worldwide concern, 24 countries meeting at Montreal in 1987 agreed to reduce production of the prime culprit, chlorofluorocarbons (CFCs), by 35 percent by 1999. CFCs are used as refrigerator and air conditioner coolants, as primary components of styrofoam, and as propellant gases in spray cans (banned in the United States in the 1970s, but still used in other countries). Whenever we buy a hamburger or a cup of coffee in a styrofoam container, whenever our automobile air conditioner leaks, or we turn in an old refrigerator for a new one, we are inadvertently contributing to upper atmosphere ozone depletion. Alternatives to CFCs are now being sought, but much work needs to be done by science, by Congress in regulating CFCs, and by all of us in changing the habits of our everyday lives.[6] These disruptions of the atmospheric balance of gases by industrial production are intimately connected to disruption of global waters.

WATER

From high mountain lakes to wild rushing rivers, the waters of the United States are threatened by acid rain. Beaches are inundated by solid wastes; globules of oil float on the surface of even the remotest oceans. Plastic wastes in the oceans are causing the deaths of upwards of 2 million birds and 100,000 marine mammals a year. Dead and dying birds entangled in plastic six-pack rings appear on beaches every day. The plastic rings will go on for another 450 years, outliving the generations they are extinguishing. Seabirds, fish, turtles, and whales lunch on small plastic pellets produced as wastes in the plastics industries. Diving birds and mammals are entrapped in plastic drift nets 6 to 30 miles in length used primarily by Japanese and other East Asian fishers. Seven hundred miles of nets are lost each season in the Pacific ocean. When these nets escape they go on trapping marine life until they sink under their own weight.[7] Global water pollution needs to be halted and water quality restored.

SOILS

Soil erosion and pollution from insecticides with long lasting half-lives are threatening croplands and ground water quality. In the United States two billion tons of topsoil is being lost annually through wind and water erosion, threatening one-third of our croplands. If allowed to continue over the next fifty years, United States' grain production will sink to about half of what it exported in 1980, affecting millions of people around the world.[8] In India, land has been used to feed people for over forty centuries, with only 5 to 10 percent of the surpluses leaving the local villages. According to conservationist Vandana Shiva, Green Revolution farming techniques have now replaced traditional methods, teaching Indian farmers "to forget about the hunger of the soil and the stomach and to go after their own hunger for profits." Soil conservation and sustainable agriculture based on the wisdom of traditional peoples need to be conbined with many of the positive advances in twentieth-century agriculture.[9]

BIOTA

Today, the reproduction of life itself is being aborted. In the words of *Time* magazine, "the death of birth" poses another immense global threat to all nonhuman species. A National Science Foundation study predicts that a quarter of the earth's species of plants, animals, microbes, and fungi will become extinct over the next several years unless extraordinary measures are taken to protect the ecosystems in which they live. Only 1.4 million of the 5 to 10 million species of life in the world have even been named. Increased efforts must be taken to identify them, understand their ecology, and to educate the public in the need for preservation.[10] International agreements have been reached on halting some of the most visible threats. The United States and Europe have recently banned imports of ivory from the African elephant. Japan has reduced imports of some endangered species such as the Hawksbill Turtle used for exotic ornaments and wedding gifts. But changes in policies and practices may not be in time to preserve the lives of known endangered species, much less those not even identified.[11]

Forests that absorb carbon dioxide and produce oxygen, linking air, water, and biota in a unity, are disappearing at a rapid rate. Tropical forests, which cover 2.3 million square miles of the earth's surface, are disappearing at the rate of 100 acres a minute or more and the rate of destruction is increasing. If the destruction continues, it is predicted that little will be left by the year 2040. The United States imports enough timber from tropical rainforests each year to cover the state of West Virginia.[12] In Central and Latin America, rainforests are being cut down to pasture cattle for the fast food industry. In Indonesia, 500,000 acres of rainforest have been converted to eucalyptus plantations to produce toilet paper for North America. Much of the rainforest being slashed in Malaysia is used by Japan to construct throwaway construction forms, boxes for shipping, and disposable chopsticks. In every inlet along the coasts of Papua New Guinea, Japanese ships anchor to receive timber, leaving behind slash as waste on beaches. Quoting Mahatma Gandhi at a June 1989 conference on "The Fate and Hope of the Earth" held in Managua, Nicaragua, Martin Khor of Indonesia admonished, "There are enough world resources for everyone's need, but not for everyone's greed."

In the United States, Pacific old-growth redwood and Douglas Fir forests are threatened by logging for export to the Far East. Seventy percent of the total harvest of uncut logs are exported—enough for 37,000 jobs in the wood products industry. Through modernization over the past decade, labor-intensive lumber mills are being replaced by automation, reducing by one-third the number of jobs available. In the process, the Spotted Owl is threatened with extinction and loggers and millers with job losses.[13] Trying to resolve complex problems such as these will require enormous sensitivity, as well as lifestyle changes on the part of northern hemisphere citizens.

Threats to the reproduction of nonhuman life are directly linked to affects on human reproduction. Toxic chemicals range from factory emissions, smog, and radon in the air, to pesticides in the soil, to trichloroethylene in drinking water. According to environmentalist Barry Commoner, humans and other living things are being invaded by an immense number of toxic chemicals unknown to biological evolution. "An organic compound," he argues, "that does not occur in nature [is] one that has been rejected in the course of evolution as incompatible with living systems." Because of their toxicity, "they have a very high probability of interfering with living processes." Over the past thirty years the production of organic chemicals from petroleum has increased from about 75 billion pounds per year to over 350 billion. In 1986 concerns such as these led California citizens to pass Proposition 65, an anti-toxics initiative with a 63 percent vote. There are presently 242 chemicals on the state's list being examined for their risk of causing cancer or birth defects.[14] Citizen actions, such as those being undertaken by the National Toxics Campaign, along with scientific research, are a vital part of the current effort to reduce toxics in the environment.

The global ecological crisis involves all levels of society—production, reproduction, and worldviews—and differentially affects First, Second, and Third World peoples.[15] The mixing and transferring of our planet's air, waters, soils, and biota that are publicized as global warming and ozone depletion are not solely the results of interacting physical, chemical, and biological systems. Such a scientific systems view ignores the linkages among processes of production, reproduction, consumption, depletion, and pollution that accompany human economies. Through commodity production and exchange, the rich soils, fossil fuels, minerals, and forests of the Third World end up in the First World as wastes in landfills and pollutants in rivers. Outlawed pesticides and toxic wastes from the First World make their way to the Third World for sale and disposal. When the price of oil rises in the Persian Gulf, First World consumers pay more at the pumps, but Third World tractors are idled and women walk an extra mile for cooking fuel. In First and Second World countries, production and consumption lead to overloaded ecological systems, while in Third World countries, resource extraction leads to exhausted and depleted lands. Economic development is uneven—centers of commerce and consumption toward which goods flow become "overdeveloped;" places on the periphery from which goods and resources flow remain "underdeveloped."[16]

The relationships between ecology and production lead to the first contradiction that constitutes the global ecological crisis. Human production systems put increasing stress on nonhuman nature through the biogeochemical cycles and energy exchanges that unify all ecological processes. As depletion and pollution accelerate, they exceed the resilience of nonhuman nature, severely undermining its capacity to recover from human-induced assaults. Systems of production, however, can be oriented toward basic subsistence, as they are in much of the Third World and indigenous cultures, or toward market exchange, as they are in First World capitalist economies and dependent Third World colonial economies. Different systems of production have different ecological impacts that result from historically different patterns of economic development.

POLITICAL ECONOMY

The patterns of uneven development and their differential economic and ecological effects are the products of a global market economy that has been emerging since the sixteenth century. The growth of a capitalist system in the European world was intimately connected to and dependent on a colonial system in the New World. As feudalism— based on the payment of goods and services to a lord by serfs bound to the land—broke down, a dynamic market system began to exploit both land and labor in more efficient ways. Mining and textile production were the first industries to be capitalized. Each expanded through the establishment of a company whose entrepreneurs pooled their wealth to take the risk of developing a mine, establishing a colony, or combining the operations of textile production under a single roof. The capitalists employed laborers who were paid in set wages from which they purchased their own food and clothing, rather than producing it from the land.

European capitalism expanded through the establishment of colonies in the western and southern hemispheres that supplied both the natural resources and cheap labor that extracted them from the earth. The former hegemony of the Mediterranean world gave way to the new hegemony of the Atlantic. Triangular trading patterns established Europe as the center of manufactured goods, Africa as the source of slave labor, and the American colonies as the "inexhaustible" supply of natural resources. The oceans were charted, the new lands mapped, and the natural histories of the peoples, animals, plants, and minerals found there catalogued. European explorers and colonizers brought with them an ecological complex of diseases that devastated native peoples and livestock, crops, weeds, and varmints that invaded native lands. The colonies were maintained by force of arms, by economic dependency on trade items, by enslavement, and by religious ideologies as missionaries worked to supplant animistic religions with Judeo-Christian theologies.[17]

Accumulation of economic surplus occurred as natural resources (or free raw materials) were extracted at minimum costs (minimum wages) and manufactured goods were sold at market value. This accumulation of economic surplus through mercantile expansion helped to fuel eighteenth and nineteenth century industrialization. Textiles and shoes, guns and ammunition, mechanized farming equipment, and standardized consumer products all depended on atomized replaceable parts and atomized replaceable laborers. Fewer people lived off the land by subsistence and more worked in cities fed by specialized market farmers. Since the period of Europe's industrial revolution (1750–1850) and North America's (1800–1900), no countries outside of those in the former Soviet bloc have been able to industrialize without economic assistance and dependency.

Today's global capitalist system is based on this same fundamental division between the industrialized or center economies of the First World and the underdeveloped or peripheral economies of the Third World. Unlike the industrialized nations, the peripheral economies export low cost primary goods such as coffee, tobacco, sugar, jute, rubber, and minerals, and import luxury goods and military equipment for élite consumption. Mass consumer goods are produced through northern hemisphere capital (Western Europe, North America, and Japan) and southern hemisphere labor (Asia, Latin America, and Africa) for purchase by northern consumers and Third World élites. Instead of enslavement by force or theft of resources, neocolonialism uses economic investments and foreign aid programs to maintain economic hegemony. Today the cost of interest on debt equals or exceeds total export earnings. The poorer countries have become increasingly dependent on the industrialized countries.

While much of the development aid to the Third World is based on First World development patterns, this undifferentiated growth model is inadequate for breaking the Third World dependency cycle. Environmental problems in the Third World are rooted in poverty and hunger, population pressure on marginal lands, and unbalanced land distribution, while those in the First World stem from industrial pollution, waste, conspicuous consumption, and planned obsolescence.[18]

A major problem confronting the capitalist system is the inherent necessity for economic growth. Capitalists make money for further expansion by creating products that consumers will purchase. They do so by fabricating needs for more and fancier food, clothing, and homes, as well as producing luxury items such as better cars, television sets, video recorders, electric shavers, blenders, and microwave ovens. Why not stop the growth mania and focus on quality of life items that fulfill basic needs? If any given producer curtails growth, she or he will be bought out or forced out of business by a competitor. If all capitalists agree together to curtail growth, massive unemployment will occur in a system in which population continues to grow.

Capitalism, however, is not isolated from government. Legislation, regulation, and citizen activism are powerful forces that can mitigate the effects of environmental pollution and improve environmental quality. Yet capitalism is historically subject to fluctuating cycles of inflation and recession and of output and unemployment. In periods of recession, concerns for environmental quality are overridden by attempts to increase productivity and employment. Governmental regulation may decline in the attempt to shore up the economic recovery. In relatively affluent periods, citizen demands for environmental quality tend to increase, as reflected in environmental movements and legal actions. Yet over time environmental quality may tend to lose ground, not returning to former levels during the peaks of relative affluence. Additionally, the environmental preferences and commitments of the political party in control of government agencies and legislatures during any given period may have positive or negative effects on the level of government regulation. All these factors are part of the structure of the social relations of the economic system of a given country and must be seen as interacting with the economy and adding to the complexity of environmental problems and their resolution.

ENVIRONMENTAL PROBLEMS IN THE SECOND WORLD

The former Soviet Union and eastern European countries are experiencing environmental problems of a different character than those of the First and Third worlds. Former president Mikhail Gorbachev's policy of *glasnost*, or openness, revealed massive amounts of industrial pollution threatening air, water, and food qualities to such an extent that citizens have become increasingly alarmed about their own health. A gas-processing plant in the city of Astrakhan pumps a million tons of sulphur into the atmosphere a year. Local people have been issued gas masks for emergency protection. In the industrial city of Nizhni Tagil, 700 miles east of Moscow, the smog is so thick that drivers turn on their headlights at noon. Throughout the commonwealth, vehicles use older engines that operate on gasoline with high lead content.

In Arkhangelsk, workers contracted diseases that were traced to the Chernobyl nuclear disaster of 1986. Although animals grazing in the area of Chernobyl were "officially" killed to prevent radiation contamination, some of the meat was transferred to remote areas and mixed with other meat to make sausages, causing the illnesses. In the cotton-producing areas of central Asia, the Aral Sea has dried to form a dustbowl. A pulp processing factory on the shores of Lake Baikal, the largest, clear fresh-water lake in the world, has created a 23 mile wide polluted area and its smoke emissons have affected 770 square miles of surrounding wilderness.[19]

In Poland, in an industrial area near Cracow, people retreat to a clinic in an underground salt mine to breathe cleaner air when smog levels are especially high. High concentrations of toxic metals such as lead, mercury, and cadmium are found in the placentas of birthing women caused by sulphur dioxide and carbon monoxide in the air. Premature births and miscarriages result from low oxygen levels in fetuses stemming from chemical changes in the mother's blood. In agricultural areas, soil is contaminated by wind and water that spread the sulphur emissons from coal burning plants over large areas. In Czechoslovakia, 50 percent of the country's drinking water does not reach minimum standards, and in Prague people complain of continual headaches, asthma, and nausea from polluted air. In eastern Germany, cancer, lung, and heart disease rates are 15 to 20 percent higher than in Berlin.[20]

Both the governments and citizens of Second World countries are taking action to curtail pollution. The former Soviet Union created a State Committee for the Protection of the Environment. Citizen groups have spearheaded conservation efforts and demonstrations against industrial polluters. Gorbachev, whose training was in agriculture, emerged as an outspoken world leader on environmental issues, and under his regime fines were levied and factories closed.

How do Second World environmental problems compare with those of the First World? Do the capitalist and socialist systems have the same environmental problems? Do economic systems matter when it comes to questions of environmental deterioration? In searching for answers, it is important to recognize both differences of kind and differences of degree. Some observers have argued that because pollution is found in both types of economies, either the problem lies in industrialization or that capitalism's problems are less severe and more easily resolved. An example of this approach was presented by economist Marshall I. Goldman in his 1970 classic paper, "The Convergence of Environmental Disruption," whose subheading encapsulated his argument: "From Lake Erie to Lake Baikal, Los Angeles to Tbilisi, the debates and dilemmas are the same." By matching cases of environmental disruption in the two countries, he drew the conclusion that they were equally polluted. His convergence thesis was as follows:

> Most conservationists and social critics are unaware that the U.S.S.R. has environmental disruption that is as extensive and severe as ours. . . . Yet before we can find a solution to the environmental disruption in our own country, it is necessary to explain why it is that a socialist or communist country like the U.S.S.R. finds itself abusing the environment in the same way, and to the same degree, that we abuse it.[21]

The United States and the former Soviet republics are all committed to economic growth. The Soviet Union and eastern European countries achieved growth through an all-out effort to raise standards of living by means of industrialization and full employment. Central government planning was the decision-making method and bureaucrats were rewarded for gross productive output. The environment suffered the consequences. Yet an important distinction exists between environmental problems in the US and the former USSR. In the Soviet Union environmental disruption stemmed largely from the effects of industrial production rather than from consumption. Packaging, plastic products, cartons, disposable diapers, styrofoam containers, household products, spray cans, aluminum soft drink cans, paints, newspapers, paper products, and other accoutrements of a disposable consumer-oriented society that choke United States' landfills and pollute its soil, air, and water are not major environmental problems in the Second World. Twenty choices of cold cereals in gaudy boxes, fifteen types of frozen diet dinners with plastic microwaveable trays, and nineteen varieties of soft drinks in nineteen different colored aluminum cans do not line the shelves of Soviet stores. Heaps of rusting automobile bodies and mountains of used tires do not adorn Soviet landscapes. Corporations and advertizing agencies do not multiply products and needs in order to compete for consumers' cash.

Yes, environmental problems exist in both the capitalist and socialist systems, but the problems are not the same for both. There is no valid convergence argument based on qualitative examples and no valid quantitative formula for comparing the relative effects of environmental disruption between the two systems. A significant structural difference does exist, however. Economic growth is inherent in capitalism; it is not essential to socialism. Both systems have historically been committed to growth; both systems have experienced bureaucratic inefficiency, poor planning, ineffective regulation, and citizen protests. It is not yet clear how the Second World will resolve its current economic and environmental crises, or how much the push to adopt market economies in the new republics will exacerbate environmental problems. Perhaps new systems will emerge from the environmental crises in the three worlds. Perhaps these syntheses will deal with environmental problems in different ways. The environmental movements in the First, Second, and Third Worlds will play important roles in the outcomes (see Chapters 7–9).

POPULATION

While the first contradiction of the global crisis emerges from the interaction between human production systems and nonhuman nature, the second contradiction arises from the interaction between production and reproduction. The impact of humans' biological reproduction on the environment is not direct, but mediated through a particular system of production (see Figure I.1). Social norms and ethical systems, as well as government policies concerning abortion, welfare, and employment, help to regulate the numbers of children born into a given society. Moreover, different modes of production support different numbers of people in particular ecological habitats. The second contradiction is thus between reproduction (both biological and social) and production. The ways in which population affects the environment must be considered within the context of biological and social reproduction and their interaction with production.

The world's population has been growing steadily during the modern era. In 1987 it reached 5 billion people and is predicted to surpass 6 billion by the year 2000. It could reach 10 to 15 billion before stabilizing sometime during the next century. Sheer numbers, however, tell only part of the story. Distribution of numbers, food, and wealth are integral to the total picture. William Keppler of the University of Alaska describes population distribution in terms of a global village:

> The present population of the world is approximately five billion people. If we could, at this very monent, shrink the earth population to a village of precisely 100 people, but all the existing human ratios remain the same, the world village would look like this:
> There would be 57 Asians, 21 Europeans, 14 western hemisphere people of both North and South America, and 8 Africans. Seventy would be non-white, 30 would be white, 70 would be non-Christian and 30 would be Christian. Fifty percent of the entire world's wealth would be in the hands of only 6 people and 5 of the 6 people would be citizens of the United States of America. Seventy percent of the population would be unable to read; 50 percent would suffer malnutrition; eight would live in substandard housing; and only one would have a university education.
> When one considers our world from such an incredibly compressed perspective the need for both tolerance and understanding in a global way becomes glaringly appearent.[22]

The population bomb, say biologists Paul and Ann Ehrlich, has now exploded. Ten thousand years ago, the world population was about five million people, but by 1650 the number had increased one hundred fold to 500 million, and by 1850 to about a billion. Since the mid-twentieth century world population has been growing by 1.7 to 2.1 percent a year, doubling about every forty years, with some nations, such as Kenya, doubling in half that time, and others, such as those in northern Europe, doubling at much slower rates. By 1990 the world growth rate had slowed from about 2.1 percent in the 1960s to about 1.8 percent in 1990, that is the doubling time increased from 33

FIGURE 1.1
POPULATION GROWTH, 1750–2100

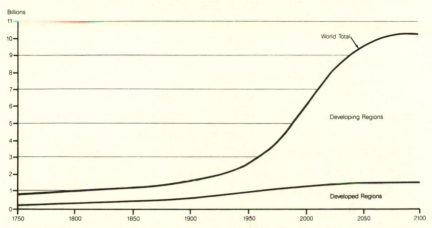

Source: Thomas W. Merrick, *et al.,* "World Population in Transition," *Population Bulletin,* Vol. 42, No. 2 (1986). Figure 1, p. 4, reprinted by permission.

to 39 years. Thus, at current rates, if the population reaches 6 billion in 2000, it will double to about 12 billion by 2040. The world would become a vast feedlot for the human species (Figure 1.1).

The Ehrlichs see all environmental problems as stemming from population: "Global warming, acid rain, depletion of the ozone layer, vulnerability to epidemics, and exhaustion of soils and groundwater are all . . . related to population size. . . . We shouldn't delude ourselves: the population explosion will come to an end before very long. The only remaining question is whether it will be halted through the humane method of birth control, or by nature wiping out the surplus."[23]

Questions of population size and control are extremely sensitive issues. They impinge on the most fundamental questions of human freedom. Freedom of how many children to bear and support, where to live, how goods and services should be distributed, a woman's right to abort a pregnancy, and the right of an unborn fetus to life. In rural China, an attempt to reduce population by a government policy of limiting families to one child resulted in the widespread abortion of female fetuses, brought about by an age-old agrarian preference for male labor. In India, Indira Gandhi's policy of pressuring sterilization of government employees after three offspring produced a backlash against its family planning program. In the United States, a woman's right to choose to abort a fetus versus the right of the fetus to life has become a major political issue in all elections, and in presidential appointments to the Supreme Court.[24]

According to the Ehrlichs, reduced fertility depends on five factors: adequate nutrition, proper sanitation, basic health care, education of women, and equal rights for women. When women receive education they apply the results to preparing better meals, keeping cleaner, more sanitary homes, and improving the quality of life for their families. Education teaches them about family planning and contraception and affords them access to status other than through bearing and raising children. Men, on the other hand, use their education to obtain higher income producing jobs, raising their status, and decreasing the need for large families. These approaches, say the Ehrlichs, rather than overall development followed by the so-called demographic transition to lower birth rates, are the keys to population control.[25]

While the interaction between population and the environment is certainly of critical concern, as are issues of women's opportunities and choices, an analysis that links all environmental problems to population growth and sees population control as the answer, say political ecologists, is too monolithic. To emphasize the impact of population on the land to the exclusion of economic development is to present a narrowly "Malthusian" perspective on the population question. In his 1798 *Essay on Population*, Thomas Malthus had argued that population tends to increase in a geometric series (2, 4, 8, 16, 64 . . .), whereas the food supply increased according to an arithmetic series (1, 2, 3, 4, 5, 6 . . .). Thus, even if the food supply could be doubled or tripled it could not keep pace with population growth. Environmental checks on population expansion, such as disease, famine, and warfare keep down the rate of increase. Rational checks such as those provided by education and foresight into the economic consequences of large families, induce birth limitation through abstinence, contraception, late marriage, and so on. Malthus argued that the educated upper-classes kept their populations down, whereas the poor reproduced at high rates. Social welfare simply encouraged them to maintain their low standard of living and their high rate of reproduction. Instead, incentives directed at individual self-interest should be provided, such as healthy work opportunites and agricultural improvement techniques.[26]

But the analysis of this "population problem" can be approached from another direction—one rooted in political economy. Geographer David Harvey argues that population, resources, and the ideologies related to their use and control must be seen in connection with economic modes of production. The number of people that a given environment can support is related to the technologies and social relations that people use to turn nonhuman nature into resources for human use.

To function at an optimal level, capitalism requires a balance between the supply of labor and the demand for goods. If the labor supply (i.e. population) increases, wages fall. Then the workers do not have enough money to buy subsistence goods. More importantly, they do not have the money to purchase commodities above the subsistence level that the capitalists wish to sell—there is no effective demand for the capitalists' products. Thus for capitalism to expand by selling more goods, wages must be kept above the subsistence level. On the other hand, if there are too few workers (i.e. a shrinking population), then wages will be too high and the capitalists will not reap sufficient profits to reinvest and expand production. For Malthus, the solution was to stimulate wants and tastes in the upper classes (landlords, state bureaucrats, etc.) thus creating fresh motives for industry. For others, such as nineteenth century economist David Ricardo, the problem could be solved by maintaining an equilibrium between capital and population, i.e. between supply and demand. Ricardo's rational, normative approach held that internal harmony within the system would allow a gradual expansion of capitalism.

A third approach is that of Karl Marx. Marx did not see a Malthusian "population problem," but a poverty and exploitation problem. Marx replaced the inevitability of the Malthusian pressure of population on the land with an analysis of the historically specific relationship between the labor supply and employment within the capitalist mode of production. Instead of the Malthusian emphasis on "overpopulation," he developed the concept of a relative surplus population. For capitalism to function smoothly, there must be a "reserve army of labor." This consists of a small percentage (about 4–5 percent)—of, for example, unemployed males, immigrants, and women,—who can be hired when the workforce shrinks and laid off when the workforce expands. In this way the capitalist can regulate both wages and demand.

When capitalists keep wages above the subsistence level, workers can purchase enough goods to maintain a reasonable quality of life. Too many children become an economic liability, rather than an asset for producing agricultural susbsistence or support for the parents in old age, keeping population growth low. If population grows too fast, however, capitalism is threatened by riots, strikes, and revolution. It thus walks a tightrope between capital, effective demand, and population. Inherent in capitalism and *essential* to its existence are abundance and scarcity, growth and natural resource depletion, and an economic division between capital and labor, i.e. between haves and have nots.[27]

Marx envisaged a society in which poverty and misery would be replaced by a system that fulfilled all people's basic needs, not just the greed of the few. Whether one agrees or disagrees with Marxist goals, a Marxist perspective offers a critical stance from which to analyze other approaches. A Marxist approach is dynamic and relational. Neither population nor resources can be understood independently of their economic context. A given part of nature is a resource or not depending on its use in a particular system. Thus gold and oil were not resources to Native Americans, but became so for European immigrants to the Americas.

Environmentalist Barry Commoner approaches population as a problem related to standards of living. The demographic transition to lower population levels is characteristic of both the industrialized world and the developing countries, but the two processes are different. As industrialization proceeded in Europe and North America, the standard of living rose and death rates declined from an average of 30 per thousand in 1850, to 24 per thousand in 1900, 16 per thousand in 1950, and 9 per thousand in 1985. Subsequently the birth rate also began to decline as fewer infants died, people lived longer, and the perceived need to bear additional children changed. The average birth rate began to decline after 1850 from 40 per thousand in 1850, to 32 per thousand in 1900, to 23 per thousand in 1950, and 14 per thousand in 1985. Overall population sizes grew during the nineteenth century, but the rate of increase slowly declined to the present rate of 0.4 percent.

In the developing countries the rate of decline has been slower. The average death rate was about 38 per thousand in 1850, 33 per thousand in 1900, 23 per thousand in 1950, and 10 per thousand in 1985. But the average birth rate has remained higher and declined much more slowly. It was 43 per thousand in 1925, 37 per thousand in 1950, and 30 per thousand in 1985. The rate of increase has slowed to about 1.7 percent a year. While death rates are about the same as those in the industrialized countries, birth rates are higher.

As the living standards improve and infant mortality declines, couples no longer need as many children to replace those who die. Instead of an economic asset to help support the parents in old age and to provide labor in agrarian communities, children become an economic liability. Costs of housing, clothing, food, travel, and a college education associated with a higher quality of life increase, providing incentives to keep family sizes smaller. Better health and childcare, better nutrition and education, steady employment, and old age security are the strongest incentives to reduction in family sizes. In addition, family planning education and safe birth control methods (as opposed to coercion and unsafe methods) provide added impetus to lowering birth rates (Figure 1.2).

In the developing countries the demographic transition has lagged because of the political and economic relationships between the center economies of the north and the peripheral economies of the south. Much of the wealth in Third World natural resources, which has been developed with northern capital and southern labor, has been removed from the southern countries. This wealth helps to fuel population decreases in the north while preventing the rise in living standards in the south that would tend to lower birth rates. The developing countries are also thwarted by enormous debts that further stall the demographic transition.

World food production is currently above the level needed to

FIGURE 1.2
AVERAGE ANNUAL RATE OF POPULATION GROWTH FOR THE WORLD, 1950–2025[a]

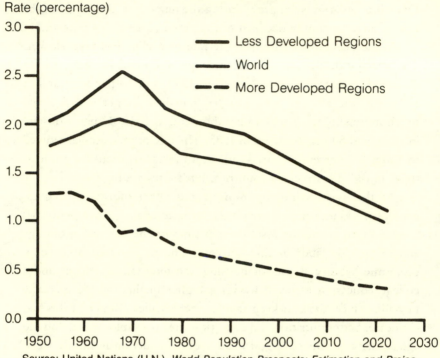

Source: United Nations (U.N.), *World Population Prospects: Estimation and Projections as Assessed in 1984* (U.N., New York, 1986), p. 25.

[a]Growth rates are based on United Nations medium variant projection of population, reprinted by permission.

support its population and the food supply is growing faster than the population. Nevertheless, that food is not evenly distributed. Some nations, such as those in Africa, have large numbers of starving people while others, such as the United States, have large food surpluses. Not only improvements in sustainable agriculture, but a redistribution of food and resources is necessary to accelerate the demographic transition.

Commoner concludes his analysis with a recommendation:

The world population crisis, which is the ultimate outcome of the exploitation of poor nations by rich ones, ought to be remedied by returning to the poor countries enough of the wealth taken from them to give their peoples both the reason and the resources voluntarily to limit their own fertility. In sum, I believe that if the root cause of the world population crisis is poverty, then to end it we must abolish poverty. And if the cause of poverty is the grossly unequal distribution of the world's wealth, then to end poverty, and with it the population crisis, we must redistribute that wealth, among nations and within them.[28]

STEADY-STATE ECONOMICS

Ultimately growth oriented economies need to move toward a steady-state world economy, argues Herman Daly. While a rapid slowdown would disproportionately affect poor countries and peoples, a gradual transition to a no- or low-growth economy could help to bring about a sustainable and socially just world. A steady-state economy, Daly says, is "an economy with constant stocks of people and artifacts, maintained at some desired, sufficient levels by low rates of maintenance 'throughput.'" The throughput is the flow of matter and energy from nonhuman nature, through the human economy, and back to nature as pollution. A steady-state economy would use the lowest possible levels of materials and energy in the production phase and emit the least possible amount of pollution in the consumption phase. The total population and the total amount of capital and consumer goods would be constant. The economy could continue to develop, but need not grow. Culture, knowledge, ethics, and quality of life would continue to grow. Only physical materials would be constant.

While the rest of the biosphere lives off solar income, human beings, since the transition to an inorganic economy, have been living off non-renewable geological capital. This means that humans are no longer in equilibrium with the rest of nature, but are depleting and polluting it, overloading the natural cycles. All capital, according to twentieth-century mathematician A. J. Lotka, is a material extension of the human body. Clothing, houses, and bathtubs are extensions of the skin; food, drink, and cooking stoves of the digestive system;

toilets and sewers of the elimination system; television and radio of the sensory organs; computers and books of the brain.

Services in the form of psychic satisfaction for humans come from increasing the numbers of artifacts and from the natural resources of the ecosystem. Creating and maintaining the artifacts requires energy throughput which in turn depletes and pollutes the ecosystem. In terms of the laws of thermodynamics, the total amount of energy in the universe is constant (the first law), but the energy available for useful work is decreasing (the second law). The total entropy (the energy unavailable for work) tends toward a maximum and the universe as a whole moves from order to disorder. As the economy uses low-entropy raw materials, it transforms them into higher-entropy arti-facts, and emits high-entropy waste. "The laws of thermodynamics," states Daly, "restrict all technologies, man's as well as nature's and apply to all economic systems whether capitalist, communist, socialist, or fascist." While the economy and its artifacts achieve greater order, the ecosystem tends to greater disorder. At some point the ecosystem will be no longer able to provide the services required by the economy. These costs to nature, however, cannot be planned in ordered se-quences as can economic costs.[29]

Is a steady-state economy possible, and if so how? Can the world of the twenty-first century move toward a stable no- or low-growth economy as population growth slows and standards of living rise? To move toward a steady-state economy, depreciation of artifacts must be reduced. Planned obsolescence gives way to planned longevity. Cars, refrigerators and television sets are engineered to last. Obsession with growth is replaced by obsession with conservation. The goal of higher gross national product gives way to the repair of gross national pollution.

CONCLUSION

Ecology, economic production, and reproduction all interact in any given society. The global ecological crisis is a result of contradictions between systems of economic production and ecology and between reproduction and production. First, Second, and Third World political economies interact in ways that exacerbate many of the problems inherent in individual countries. The political economy of the First World is legitimated by a mechanistic worldview that has been dominant since the seventeenth century and an egocentric ethic that assumes that what is best for the individual is best for society as a whole. These issues are discussed in Chapters 2 and 3. Many observers believe that the world is moving toward some new state of affairs that will radically change current patterns at all social levels. Part II on radical ecological thought and Part III on radical environmental movements put forward some ideas for transformation that may help to resolve the global crisis by attacking the contradictions that lead to it. Such changes would alter current ecological, economic, and social relations with nonhuman nature, as well as the mechanistic worldview, helping to create a sustainable world.

FURTHER READING

Amin, Samir. *Eurocentrism*. New York: Monthly Review Press, 1989.

———. *Unequal Development: An Essay on the Social Formations of Peripheral Capitalism.* New York: Monthly Review Press, 1976.

Brown, Lester. *State of the World*. New York: W. W. Norton, 1990.

Commoner, Barry. *The Closing Circle: Nature, Man and Technology*. New York: Knopf, 1971.

———. *The Poverty of Power: Energy and the Economic Crisis*. New York: Knopf, 1976.

———. *Making Peace with the Planet*. New York: Pantheon, 1990.

Crosby, Alfred. *Ecological Imperialism: The Biological Expansion of Europe, 900–1900.* New York: Cambridge University Press, 1986.

Daly, Herman. *Steady State Economics*. San Francisco: W. H. Freeman, 1977.

———. ed. *Economics, Ecology, Ethics: Essays Toward a Steady-State Economy*. San Francisco: W. H. Freeman, 1980.

Daly, Herman, and John Cobb, Jr. *For the Common Good: Redirecting the Economy Toward Community, the Environment, and a Sustainable Future*. Boston: Beacon Press, 1989.

Edwards, Richard C. et. al., eds. *The Capitalist System*. London: Prentice Hall, 1972.

Ehrlich, Paul. *The Population Bomb*. New York: Ballantine Books, 1968.

Ehrlich, Paul R., and Ann H. Ehrlich. *The Population Explosion*. New York: Simon and Schuster, 1990.

George, Susan. *A Fate Worse Than Debt: The World Financial Crisis and the Poor*. New York: Grove Press, 1988.

Goldfarb, Theodore. *Taking Sides: Clashing Views on Controversial Environmental Issues*. Guilford, Ct.: Dushkin Publishing Group, 1989.

Meek, Ronald, ed. *Marx and Engels on the Population Bomb*. Berkeley, Ca.: Ramparts Press, 1971.

Meeker-Lowry, Susan. *Economics as If the Earth Really Mattered*. Philadelphia, Pa.: New Society Publishers, 1988.

Merchant, Carolyn. *Ecological Revolutions: Nature, Gender, and Science in New England*. Chapel Hill: University of North Carolina Press, 1989.

Miller, Alan. *A Planet to Choose: Value Studies in Political Ecology*. New York: Pilgrim Press, 1978.

Myers, Norman, ed. *Gaia: An Atlas of Planet Management*. Garden City, N.Y.: Anchor Doubleday, 1984.

Polanyi, Karl. *The Great Transformation: The Political and Economic Origins of Our Time*. New York: Holt, 1944.

Rifkin, Jeremy. *Entropy: A New World View*. New York: Viking Press, 1980.

World Resources Institute. *World Resources 1988–89: An Assessment of the Resource Base that Supports the Global Economy*. New York: Basic Books, 1988.

2

SCIENCE AND WORLDVIEWS

Is the earth dead or alive? The ancient cultures of east and west and the native peoples of America saw the earth as a mother, alive, active, and responsive to human action. Greeks and Renaissance Europeans conceptualized the cosmos as a living organism, with a body, soul, and spirit, and the earth as a nurturing mother with respiratory, circulatory, reproductive, and elimination systems. The relationship between most peoples and the earth was an I-thou ethic of propitiation to be made before damming a brook, cutting a tree, or sinking a mine shaft. Yet for the past three hundred years, western mechanistic science and capitalism have viewed the earth as dead and inert, manipulable from outside, and exploitable for profits. The death of nature legitimated its domination. Colonial extractions of resources combined with industrial pollution and depletion have today pushed the whole earth to the brink of ecological destruction.

THE ORGANIC WORLDVIEW

The cosmos of the Renaissance world was a living organism. The four elements (earth, air, fire, and water) that made up the material world below the moon, and the fifth element (ether) that made the stars and planets were its material body. The soul was the source of its animate daily motion as the sun, stars, and planets encircled the geocentric earth every twenty-four hours. The spirit, descending from God in the heavens beyond, mingled with the ether and the ambient air, to be imbibed by plants, animals, and humans on the earth's surface.

The living character of the world organism meant not only that the stars and planets were alive, but that the earth too was pervaded by a force giving life and motion to the living beings on it. The earth was considered to be a beneficient, receptive, nurturing female. In the ancient lore, the earth mother respired daily, inhaling the pneuma, or spirit from the atmosphere. Her "copious breathing" renewed the life on its surface. The earth's springs were akin to the human blood system; its other various fluids were likened to the mucus, saliva, sweat, and other forms of lubrication in the human body. As the waters on its surface ebbed and flowed, evaporated into clouds, and descended as dews, rains, and snows, the earth's blood was cleansed and renewed. Veins, veinlets, seams, and canals coursed through the entire earth, particularly in the mountains. It humors flowed from the veinlets into larger veins. In many places the veins became filled with metals and minerals.

The earth, like the human, even had its own elimination system. The tendency for the earth to break wind was the cause of earthquakes and a manifestation of the earth mother's indignation at humans who mined her entrails. The earth's bowels were full of channels, fire chambers, glory holes, and fissures through which fire and heat were emitted, some in the form of fiery volcanic exhalations, other as hot water springs. The thin layer of soil on the earth's surface was its skin. European peasants nurtured the land, performed ritual dances, and returned its gifts to assure continued fertility. Trees were the earth mother's tresses. Her head was adorned with fringes and curls which the lumber industry sheared off.

A commonly used analogy was that between the female's reproductive and nurturing capacity and the mother earth's ability to give birth to stones and metals within "her" womb through marriage with the sun. For most traditional cultures, minerals and metals ripened in the uterus of the Earth Mother, mines were compared to her vagina, and metallurgy was the human hastening of the birth of the living metal in the artificial womb of the furnace—an abortion of the metal's natural growth cycle before its time. Miners offered propitiation to the deities of the soil and subterranean world, performed ceremonial sacrifices, and observed strict cleanliness, sexual abstinence, and fasting before violating the sacredness of the living earth by sinking a mine. Smiths assumed an awesome responsibility in precipitating the metal's birth through smelting, fusing, and beating it with hammer and anvil; they were often accorded the status of shaman in tribal rituals, and their tools were thought to hold special powers.

The image of the earth as a living organism and nurturing mother served as a cultural constraint restricting the actions of human beings. One does not readily slay a mother, dig into her entrails for gold, or mutilate her body. As long as the earth was conceptualized as alive and sensitive, it could be considered a breach of human ethical behavior to carry out destructive acts against it. In much the same way, the cultural belief-systems of many American Indian tribes had for centuries subtly guided group behavior toward nature. Smohalla of the Columbian Basin Tribes voiced the Indian objections to European attitudes in the mid-1800s.

> You ask me to plow the ground! Shall I take a knife and tear my mother's breast? Then when I die she will not take me to her bosom to rest.
> You ask me to dig for stone! Shall I dig under her skin for her bones? Then when I die I cannot enter her body to be born again.

You ask me to cut grass and make hay and sell it, and be rich like white men! But how dare I cut off my mother's hair?

Such imagery found in a culture's literature can play a normative role within the culture. Controlling images operate as ethical restraints or as ethical sanctions—as subtle "oughts" or "ought-nots." Thus, as the descriptive metaphors and images of nature change, a behavioral restraint can be changed into a sanction. Such a change in the image and description of nature was occurring during the course of the scientific revolution. Today, the organic cosmology, experienced in some form by almost all of the world's peoples for all times, has been superseded.[1]

THE RISE OF CAPITALISM

In the sixteenth century, as the feudal states of medieval Europe were breaking up, a new dynamic force emerged that shattered premodern ways of life and the organic restraints against the exploitation of the earth. Arising in the city-states of Renaissance Italy and spreading to northern Europe was an inexorable expanding market economy, intensifying medieval tendencies toward capitalist relations of production and capitalist modes of economic behavior. As trade quickened throughout western Europe, stimulated by the European discovery and exploitation of the Americas, production for subsistence began to be replaced by more specialized production for the market. The spreading use of money provided not only a uniform medium of exchange but also a reliable store of value, facilitating open-ended accumulation. Inflation generated by the growth of population and the flood of American gold accelerated the transition from traditional economic modes to rationally maximizing modes of economic organization. The growth of cities as centers of trade and handicraft production created a new class of bourgeois entrepreneurs who supplied ambitious monarchs with the funds and expertise to build strong nation states, undercutting the power of the regionally based landowning nobility.

Whereas the medieval economy had been based on organic and renewable energy sources—wood, water, wind, and animal muscle—the emerging capitalist economy was based on nonrenewable energy—coal—and the inorganic metals—iron, copper, silver, gold, tin, and mercury—the refining and processing of which ultimately depended on and further depleted the forests. Over the course of the sixteenth century, mining operations quadrupled as the trading of metals expanded, taking immense toll as forests were cut for charcoal and the cleared lands turned into sheep pastures for the textile industry. Shipbuilding, essential to capitalist trade and national supremacy, along with glass and soap-making, also contributed to the denudation of the ancient forest cover. The new activities directly altered the earth. Not only were its forests cut down, but swamps were drained, and mine shafts were sunk.

The new commercial and industrial enterprises meant that the older cultural constraints against the exploitation of the earth no longer held sway. While the organic framework was for many centuries sufficiently integrative to override commercial development and technological innovation, the acceleration of economic change throughout western Europe began to undermine the organic unity of the cosmos and society. Because the needs and purposes of society as a whole were changing with the commercial revolution, the values associated with the organic view of nature were no longer applicable; hence the plausibility of the conceptual framework itself was slowly, but continuously, being threatened. By the sixteenth and seventeenth centuries, the tension between the technological development in the world of action and the controlling organic images in the world of the mind had become too great. The old worldview was incompatible with the new activities.[2]

EXPERIMENTAL SCIENCE

During the seventeenth century, the organic framework, in which the Mother-Earth image was a moral restraint against the exploitation of nature, was replaced by a new experimental science and a worldview that saw nature not as an organism but as a machine—dead, inert, and insensitive to human action. Francis Bacon (1571–1626), following tendencies that had been evolving throughout the previous century, advocated the domination of nature for human benefit. He compared miners and smiths whose technologies extracted ores for the new commercial activities to scientists and technologists penetrating the earth and shaping "her" on the anvil. The new man of science, he wrote, must not think that the "inquisition of nature is in any part interdicted or forbidden." Nature must be "bound into service" and made a "slave," put "in constraint," and "molded" by the mechanical arts. The "searchers and spies of nature" were to discover her plots and secrets.[3]

Nature's womb, Bacon argued, harbored secrets that through technology could be wrested from her grasp for use in the improvement of the human condition. Before the fall of Adam and Eve there had been no need for power or dominion, because they had been made sovereign over all other creatures. Only by "digging further and further into the mine of natural knowledge," Bacon believed, could mankind recover that lost dominion. Nature placed in bondage through technology would serve human beings. Here "nature takes orders from man and works under his authority." The method of science was not to be achieved by developing abstract notions such as those of the medieval scholastics, but rather through the instruction of the understanding "that it may in very truth dissect nature." "By art and the hand of man," nature should be "forced out of her natural state and squeezed and molded." In this way "human knowledge and human power meet as one."[4]

Thus Bacon, in bold sexual imagery, outlined the key features of the modern experimental method—constraint of nature in the laboratory, dissection by hand and mind, and the penetration of nature's hidden secrets—language still used today in praising a scientist's "hard facts," "penetrating mind," or "seminal" arguments. The constraints against mining the earth were subtly turned into sanctions for exploiting and "raping" nature for human good.[5]

The development of science as a methodology for manipulating nature, and the interest of scientists in the mechanical arts, became a significant program during the latter half of the seventeenth century.

Other philosophers realized even more clearly than had Bacon himself the connections between mechanics, the trades, middle-class commercial interests, and the domination of nature. Scientists spoke out in favor of "mastering" and "managing" the earth. French Philosopher René Descartes wrote in his *Discourse on Method* (1637) that through knowing the crafts of the artisans and the forces of bodies we could "render ourselves the masters and possessors of nature."[6]

John Dury and Samuel Hartlib, English Baconians and organizers of the Invisible College (ca. 1645), connected the study of the crafts and trades to increasing wealth. The members of England's first scientific society, the Royal Society (founded in 1660), were interested in carrying out Bacon's proposals to dominate nature through experimentation. Joseph Glanvill, the English philosopher who defended the Baconian program in his *Plus Ultra* (1668), asserted that the objective of natural philosophy was to "enlarge knowledge by observation and experiment . . . so that nature being known, it may be mastered, managed, and used in the services of humane life." For Glanvill, anatomy, was "most useful in human life" because it "tend[ed] mightily to the eviscerating of nature, and disclosure of the springs of its motion." In searching out the secrets of nature, nothing was more helpful than the microscope for "the secrets of nature are not in the greater masses, but in those little threads and springs which are too subtle for the grossness of our unhelped senses."[7]

In his *Experimental Essays* (1661), English scientist Robert Boyle distinguished between merely knowing as opposed to dominating nature in thinly veiled sexual metaphor: "For some men care only to know nature, others desire to command her" and "to bring nature to be serviceable to their particular ends, whether of health, or riches, or sensual delight."[8]

The experimental method developed by the seventeenth-century scientists was strengthened by the rise of the mechanical philosophy. Together they replaced the older, "natural" ways of thinking with a new and "unnatural" way of seeing, thinking, and behaving. The submergence of the organism by the machine engaged the best minds of the times during a period fraught with anxiety, confusion, and instability in both the intellectual and social spheres.

THE MECHANISTIC WORLDVIEW

The mechanical view of nature now taught in most western schools is accepted without question as our everyday, common sense reality—a reality in which matter is made up of atoms, colors occur by the reflection of light waves of differing lengths, bodies obey the law of inertia, and the sun is in the center of our solar system. This worldview is a product of the scientific revolution of the seventeenth century. None of its assumptions were the commonsense view of our sixteenth-century counterparts. Before the scientific revolution, most ordinary people assumed that the earth was in the center of the cosmos, that the earth was a nurturing mother, and that the cosmos was alive, not dead.

As the unifying model for science and society, the machine has permeated and reconstructed human consciousness so totally that today we scarcely question its validity. Nature, society, and the human body are composed of interchangeable atomized parts that can be repaired or replaced from outside. The "technological fix" mends an ecological malfunction, new human beings replace the old to maintain the smooth functioning of industry and bureaucracy, and intervention-ist medicine exchanges a fresh heart for a worn-out, diseased one.

The removal of animistic, organic assumptions about the cosmos constituted the death of nature—the most far-reaching effect of the scientific revolution. Because nature was now viewed as a system of dead, inert particles moved by external rather than inherent forces, the mechanical framework itself could legitimate the manipulation of nature. Moreover, as a conceptual framework, the mechanical order had associated with it a framework of values based on power, fully compatible with the directions taken by commercial capitalism.[9]

The emerging mechanical worldview was based on assumptions about nature consistent with the certainty of physical laws and the symbolic power of machines. Although many alternative philosophies were available (Aristotelian, Stoic, gnostic, Hermetic, magic, natural-ist, and animist), the dominant European ideology came to be governed by the characteristics and experiential power of the machine. Social values and realities subtly guided the choices and paths to truth and certainty taken by European philosophers. Clocks and other early modern machines in the seventeenth century became underlying models for western philosophy and science.

Not only were seventeenth-century philosophical assumptions about being and knowledge infused by the fundamental physical structures of machines found in the daily experience of western Europeans, but these presuppositions were completely consistent with another feature of the machine—the possibility of controlling and dominating nature. These underlying assumptions about the nature of reality have today become guidelines for decision-making in technology, industry, and government.

The following assumptions about the structure of being, knowledge, and method make possible the human manipulation and control of nature.

1. Matter is composed of particles (the ontological assumption).
2. The universe is a natural order (the principle of identity).
3. Knowledge and information can be abstracted from the natural world (the assumption of context independence).
4. Problems can be analyzed into parts that can be manipulated by mathematics (the methodological assumption).
5. Sense data are discrete (the epistemological assumption).[10]

The new conception of reality developed in the mid-seventeenth century shared a number of assumptions with the clocks, geared mills, and force-multiplying machines that had become an important part of daily European economic life. First of all, they shared the ontological assumption that nature is made up of modular components or discrete parts connected in a causal nexus that transmitted motion in a temporal sequence from part to part. Corpuscular and atomic theories revived in the seventeenth century hypothesized a particulate structure to reality. The parts of matter, like the parts of machines, were dead, passive, and inert. The random motions of atoms were rearranged to form new objects and forms of being by the action of external forces. Motion was not inherent in the corpuscles, but a primary quality of matter, put into the mundane machine by God. In Descartes' philosophy, motion was initiated at the world's creation and sustained from instant to instant throughout created time; for English physicist Isaac Newton (1642–1727), new motion in the form of "active principles" (the cause of gravity, fermentation, and electricity) was added periodically to prevent the nonautonomous world-machine from running down. For German philosopher Gottfried Wilhelm Leibniz (1646–1716), the universal clock was autonomous—it needed no external inputs once created and set into motion. The ontology of this classical seventeenth-century science, modified by energy concepts, has become the framework of the western commonsense view of reality.

The second shared assumption between machines and seventeenth-century science was the law of identity, the idea that A is A, or of identity through change. This assumption of a rational order in nature goes back to the thought of the philosophers Parmenides of Elea (fl. 500 B. C.) and Plato (4th century B. C.) and is the substance of Aristotle's first principle of logic. Broadly speaking, it is the assumption that nature is subject to lawlike behavior and therefore that the domain of science and technology includes those phenomena that can be reduced to orderly predictable rules, regulations, and laws. Events that can be so described can be controlled because of the simple identity of mathematical relationships. Phenomena that "cannot be foreseen or reproduced at will . . . [are] essentially beyond the control of science."[11]

The formal structural dependence of this mathematical method on the features of the mechanical arts was beautifully articulated by Descartes in his *Discourse on Method* (1636): "Most of all I was delighted with mathematics, because of the certainty of its demonstrations and the evidence of its reasoning; but I did not understand its true use, and, believing that it was of service only in the mechanical arts, I was astonished that, seeing how firm and solid was its basis, no loftier edifice had been reared thereupon."[12]

The primary example of the law of identity for Descartes was conservation of the quantity of motion measured by the quantity of matter and its speed, $m|v|$. In the late-seventeenth century Newton, Leibniz, English mathematicians Christopher Wren and John Wallis, and Dutch physicist Christiaan Huygens all contributed to the correction of Descartes' law accurately to describe momentum (mv) as the product of mass and velocity rather than speed, and mechanical energy (mv^2) as the product of the mass and the square of the velocity. Everyday machines were models of ideal machines governed and described by the laws of statics and the relational laws of the conservation of mechanical energy and momentum. The form or structure of these laws, based as they were on the law of identity, was thus a model of the universe. Although the conversion of energy from one form to another and, in particular, the conversion of mechanical motion into heat were not fully understood until the nineteenth century, the seventeenth-century laws of impact were nevertheless, for most natural philosophers, models of the transfer and conservation of motion hypothesized to exist in the ideal world of atoms and corpuscles.

The third assumption, context independence, goes back to Plato's insight that only quantities and context-independent entities can be submitted to mathematical modeling. To the extent that the changing imperfect world of everyday life partakes of the ideal world, it can be described, predicted, and controlled by science just as the physical machine can be controlled by its human operator. Science depends on a rigid, limited, and restrictive structural reality. This limited view of reality is nevertheless very powerful, inasmuch as it allows for the possibility of control whenever phenomena are predictable, regular, and subject to rules and laws. The assumption of order is thus fundamental to the concept of power, and both are integral to the modern scientific worldview.[13]

Although Descartes' plan for reducing complexity in the universe to a structured order was comprehensive, he discovered that the very problem that Aristotle had perceived in the method of Plato was inherent in his own scheme. That problem was the intrinsic difficulty, if not impossibility, of successfully abstracting the form or structure of reality from the tangled web of its physical, material, environmental context. Structures are in fact not independent of their contexts, as this third assumption stated, but integrally tied to them. In fact, Descartes was forced to admit, "the application of the laws of motion is difficult, because each body is touched by several others at the same time. . . . The rules presuppose that bodies are perfectly hard and separable from all others. . .and we do not observe this in the world." The enormous complexity of things thus inhibits the analysis in terms of simple elements.[14]

Descartes' method exhibits very precisely the fourth or methodological assumption that problems can be broken down into parts and information can then be manipulated in accordance with a set of mathematical rules and relations. Succinctly stated, his method assumes that a problem can be analyzed into parts, and that the parts can be simplified by abstracting them from the complicating environmental context and then manipulated under the guidance of a set of rules.

His method consisted of four logical precepts:

1. To accept as true only what was so clearly and distinctly presented that there was no reason to doubt it;
2. To divide every problem into as many parts as needed to resolve it;
3. To begin with objects simple and easy to understand and to rise by degrees to the most complex);
4. To make so general and complete a review that nothing is omitted.

In Descartes' opinion, this method was the key to power over nature, for these methods of reasoning used by the geometricians "caused me to imagine that all those things which fall under the cognizance of man might very likely be mutually related in the same fashion." By following this method, "there can be nothing so remote that we cannot reach to it, or recondite we cannot discover it."

Descartes' method depended on the manipulation of information according to a set of rules: "Commencing with the most simple and general (precepts), and making each truth that I discovered a rule for helping me to find others,—not only did I arrive at the solution of many questions which I had hitherto regarded as most difficult but ... in how far, it was possible to solve them." In the same manner, the operation of a machine depends on the manipulation of its material parts in accordance with a prescribed set of physical operations.

Descartes placed great emphasis on the concept of a plan or form for ordering this information, drawing his examples from the practical problem of city planning: "Those ancient cities which, originally mere villages, have become in the process of time great towns, are usually badly constructed in comparison with those which are regularly laid out on a plain by a surveyor who is free to follow his own ideas." He wished his new ideas to "conform to the uniformity of a rational scheme."[15]

In his *De Cive,* written in 1642, Hobbes had advocated the application of this method of analysis to society:

> For everything is best understood by its constitutive causes. For as in a watch, or some such small engine, the matter, figure, and motion of the wheels cannot well be known except it be taken asunder and viewed in parts; so to make a more curious search into the rights of states and duties of subjects, it is necessary, I say, not to take them asunder, but yet that they be so considered as if they were dissolved.[16]

The fifth assumption shared by seventeenth-century science and the technology of machines was the assumption that sense data are atomic. Data are received by the senses as minute particles of information. This assumption about how knowledge is received was articulated most explicitly by Hobbes and the British empiricists John Locke and David Hume. According to Hobbes, sense data arise from the motions of matter as it affects our sense organs, directly in the case of taste and touch, or indirectly, through a material medium, as in sight, sound, and smell. These sense data can then be manipulated and recombined according to the rules of free speech: "But the most noble and profitable invention of all other, was that of speech, consisting of names or appellations and their connection whereby men register their thoughts . . . without which there had been among men neither commonwealth, nor society, nor contract, nor peace."[6] Words are abstractions from reality; sentences or thoughts are connections among words: "The manner how speech serves to the remembrance of the consequence of causes and effects, consists in the imposing of names and the connection of them." Nature cannot be understood unless it is first analyzed into parts from which information can be extracted as sense data: "No man therefore can conceive anything, but he must conceive it in some place and endowed with some determinate magnitude; and which may be divided into parts."[17]

For Hobbes, the mind itself is a special kind of a machine—a calculating machine similar to those constructed by Scottish mathematician John Napier (1550–1617), French philosopher and mathematician Blaise Pascal (1623–1662), Leibniz, and other seventeenth-century scientists. To reason is but to add and subtract or to calculate. "When a man reasoneth, he does nothing else but conceive a sum total, from the addition of parcels; or conceive a remainder, from subtraction of one sum from another; which, if it be done by words, is conceiving of the consequence of the names of all the parts, to the name of the whole; or from the names of the whole and one part to the name of the other part." "In sum, in what matter soever there is place for addition and subtraction, there is also place for reason; and where these have no place, there reason has nothing at all to do. . . . For reason . . . is nothing but reckoning, that is adding and subtracting."[18] This view is manifested in twentieth-century information theory that, according to philosopher Martin Heidegger, is "already the arrangement whereby all objects are put in such form, as to assure man's domination over the entire earth and even the planets."[19]

The new definition of reality of seventeenth-century philosophy and science was therefore consistent with, and analogous to, the structure of machines. Machines (1) are made up of parts, (2) give particulate information about the world, (3) are based on order and regularity, (perform operations in an ordered sequence), (4) operate in a limited, precisely defined domain of the total context, and (5) give us power over nature. In turn, the mechanical structure of reality (1) is made up of atomic parts, (2) consists of discrete information bits extracted from the world, (3) is assumed to operate according to laws and rules, (4) is based on context-free abstraction from the changing complex world of appearance, and (5) is defined so as to give us maximum capability for manipulation and control over nature.[20]

THE DOMINATION OF NATURE

Based on these five assumptions about the nature of reality, science since the seventeenth century has been widely considered to be objective, value-free, context-free knowledge of the external world. Additionally, as Heidegger argued, western philosophy since Descartes has been fundamentally concerned with power. "The essence of modern technology lies in enframing;" that is, in the revealing of nature so as to render it a "standing reserve," or storehouse. "Physics, indeed as pure theory," he wrote, "sets up nature to exhibit itself" in such a way as to "entrap" it "as a calculable order of forces."[21]

Both order and power are integral components of the mechanical view of nature. Both the need for a new social and intellectual order and new values of human and machine power, combined with older intellectual traditions, went into the restructuring of reality around the metaphor of the machine. The new metaphor reintegrated the disparate elements of the self, society, and the cosmos torn asunder by the Protestant Reformation, the rise of commercial capitalism, and the early discoveries of the new science.

The domination of nature depends equally on the human as operator, deriving from an emphasis on power and on the human as manager, deriving from the stress on order and rationality as criteria for progress and development. Efficient operation results from the ordered rational arrangement of the components of a system. The mechanical framework with its associated values of power and control sanctioned the management of both nature and society. The management of natural resources depends on surveying the status of existing resources, and efficiently planning their systematic use and replenishment for the long-term good of those who use them.[22]

NEWTONIAN SCIENCE

The world in which we live today was bequeathed to us by Isaac Newton. Twentieth-century advances in relativity and quantum theory notwithstanding, our western commonsense reality is the world of classical physics. The legacy left by Newton was the brilliant synthesis of Galilean terrestrial mechanics and Copernican-Keplerian astronomy. Fundamental in generality, it describes and extends over the entire universe. Classical physics and its philosophy structure our consciousness to believe in a world composed of atomic parts, of inert bodies moving with uniform velocity unless forced by another body to deviate from their straight-line paths, of objects seen by reflected light of varying frequencies, and of matter in motion responsible for all the rich variations in colors, sounds, smells, tastes, and touches we cherish as human beings. In our daily lives, most of us accept these teachings as givens, without much critical reflection on their origins or associated values.

The problem that the mechanization of the world raised for the generation after Descartes and Hobbes was the very issue of the "death of nature." If the ultimate principles were matter and motion—as they were for the first generation of mechanists—or even matter, motion, void space, and force—as they became for Newton—this left unresolved the central issue of explaining the motion of life-forms in a dead cosmos. Like many others, Newton was not satisfied with Descartes' dualistic solution, which reduced the human being to a ghost-in-the-machine whose mind could change the direction of but not initiate bodily motion, and categorized animals as mere beast machines.

Yet as the most powerful synthesis of the new mechanical philosophy, Newton's *Mathematical Principles of Natural Philosophy*, (1687) epitomized the dead world resulting from mechanism. Throughout the complex evolution of his thought, Newton clung tenaciously to the distinguishing feature of mechanism—the dualism between the passivity of matter and the externality of force and activity.[23]

Mechanism eliminated from the description of nature concepts of spatial hierarchy, value, purpose, harmony, quality, and form central to the older organic description of nature, leaving material and efficient causes—matter and force. Motion was not an organic process but a temporary state of a body's existence relative to the motion or rest of other bodies. The mathematizing tendencies in Newtonian thought which emphasized not the process of change, but resistance to change, the conservation of a body's motion, and the planets and satellites as ideal spheres and point sources of gravitational force were manifestations of the mechanical philosophers' concern with geometrical idealization, stability, structure, being, and identity, rather than organic flux, change, becoming, and process. In mechanism the primacy of process was thus superseded by the stability of structure.

Completely consistent with this restructuring of the cosmos as passive matter and external force was the division of matter into atomic parts separated by void space. The book of nature was no longer written in symbols, signs, and signatures, but in corpuscular characters. The atomic analysis of matter ultimately became an exemplar for the atomic division of data, problems, and events on a global scale.[24]

Newton's speculations on atomic structure as presented in the 1713 edition of the *Principia* and the queries to the 1706 and 1717 editions of the *Opticks* became a foundation for eighteenth-century experimental philosophers, who wished to complete the task of reducing known phenomena to simple laws which—like the law of gravitation—would quantify other mechanical, chemical, electrical, and thermal observations. Moreover, its conceptual framework, emphasizing external force and passive matter divided into rearrangeable components, could provide a subtle sanction for the domination and manipulation of nature necessary to progressive economic development. If eventually the religious framework providing for God's constant care and for the attainment of human grace were removed, as it was in the eighteenth century, the possibilities for intellectual arrogance toward nature would be strengthened.[25]

The mechanistic view of nature, developed by the seventeenth-century natural philosophers and based on a western mathematical tradition going back to Plato, is still dominant in science today. This view assumes that nature can be divided into parts and that the parts can be rearranged to create other species of being. "Facts" or information bits can be extracted from the environmental context and rearranged according to a set of rules based on logical and mathematical operations. The results can then be tested and verified by resubmitting them to nature, the ultimate judge of their validity.

Twentieth-century logical positivism, the basis for scientific knowledge, assumes that only two types of statements lead to truths about the natural world: mathematical (or logical statements) of the form $a = a$, and empirically verifiable statements. Mathematical formalism provides the criterion for rationality and certainty, nature the criterion for empirical validity and acceptance or rejection of the theory. Natural science has thus become the model for knowledge.

The mechanical approach to nature is as fundamental to twentieth-century physics as it was to classical Newtonian science. Twentieth-century physics still views the world in terms of fundamental particles—electrons, protons, neutrons, mesons, muons, pions, *taus*, *thetas*, *sigmas*, *pis*, and so on. The search for the ultimate unifying particle, the quark, continues to engage the efforts of the best theoretical physicists.

Modern science is widely assumed to be objective, value-free, context-free knowledge of the external world. The greater the extent to which the sciences can be reduced to this mechanistic mathematical model, the more legitimate they become as sciences. Thus the reductionist hierarchy of the validity of the sciences first proposed in the nineteenth century by French positivist philosopher August Comte is still widely assumed by intellectuals, the most mathematical and highly theoretical sciences occupying the most revered position.[26]

CONCLUSION

Between 1500 and 1700 an incredible transformation took place. A "natural" point of view about the world in which bodies did not move unless activated, either by an inherent organic mover or a "contrary to nature" superimposed "force," was replaced by a non-natural non-experiential "law" that bodies move uniformly unless hindered. The "natural" perception of a geocentric earth in a finite cosmos was superseded by the "non-natural" commonsense "fact" of a heliocentric infinite universe. A subsistence economy in which resources, goods, money, or labor were exchanged for commodities was replaced in many areas by the open-ended accumulation of profits in an international market. Living animate nature died, while dead inanimate money was endowed with life. Increasingly capital and the market assumed the organic attributes of growth, strength, activity, pregnancy, weakness, decay, and collapse, obscuring and mystifying the new underlying social relations of production and reproduction that made economic growth and progress possible. Nature, women, blacks, and wage laborers were set on a path toward a new status as "natural" and as human resources for the modern world system. Perhaps the ultimate irony in these transformations was the new name given them: rationality.[27]

Although the mechanistic analysis of reality has dominated the western world since the seventeenth century, the organismic perspective has by no means disappeared. It has remained as an important underlying tension, surfacing in such variations as Romanticism, American transcendentalism, the German Nature philosophers, and the early philosophy of Karl Marx. The basic tenets of the organic view of nature have reappeared in the twentieth century in the theory of holism of Jan Christiaan Smuts, the process philosophy of Alfred North Whitehead, the ecology movement of the 1970s, and David Bohm's holomovement (see Chapters 3 and 4). Some philosophers have argued that the two frameworks are fundamentally incommensurable. Others argue that a reassessment of the underlying metaphysics and values historically associated with the mechanistic worldview may be essential for a viable future.[28]

The mechanistic worldview continues today as the legitimating ideology of industrial capitalism and its inherent ethic of the domination of nature. Mechanistic thinking and industrial capitalism lie at the root of many of the environmental problems discussed in Chapter 1. The egocentric ethic associated with this worldview, however, has been challenged by the ecocentric ethic of the ecology movement (see Chapter 3) and the worldview itself by deep ecology (see Chapter 4).

FURTHER READING

Berman, Morris. *The Reenchantment of the World*. Ithaca: Cornell University Press, 1981.

Easlea, Brian. *Witch Hunting, Magic and the New Philosophy: An Introduction to Debates of the Scientific Revolution, 1450–1750*. Sussex: The Harvester Press, 1980.

———. *Science and Sexual Oppression: Patriarchy's Confrontation with Woman and Nature*. London: Weidenfeld and Nicolson, 1981.

————. *Fathering the Unthinkable: Masculinity, Scientists, and the Nuclear Arms Race.* London: Pluto Press, 1983.

Keller, Evelyn Fox. *Reflections on Gender and Science.* New Haven, Ct.: Yale University Press, 1985.

Kubrin, David. "How Sir Isaac Newton Helped Restore Law 'n Order to the West." *Liberation Magazine* 16, no. 10 (March 1972): 32–41.

Merchant, Carolyn. *The Death of Nature: Women, Ecology, and the Scientific Revolution.* San Francisco: Harper and Row, 1980.

Schiebinger, Londa. *The Mind Has No Sex? Women in the Origins of Modern Science.* Cambridge, Ma.: Harvard University Press, 1989.

3

ENVIRONMENTAL ETHICS AND POLITICAL CONFLICT

In his *Nichomachean Ethics,* Aristotle noted that "all knowledge and every pursuit aims at some good."[1] But whether this is an individual, social, or environmental good lies at the basis of many real world ethical dilemmas. Egocentric, homocentric, and ecocentric ethics often underlie the political positions of various interest groups engaged in struggles over land and natural resource uses. These ethics are the culmination of sets of associated political, religious, and ethical trends developing in western culture since the seventeenth century. Conflicts of interest among private individuals, corporations, government agencies, and environmentalists often reflect variations of these three ethical approaches. Thinking about environmental problems in terms of this taxonomy helps us to understand the unexpressed assumptions behind political conflicts over the environment.

These ethical differences are also at the root of some of the disagreements among radical environmental theorists and activists detailed in subsequent chapters. An egocentric ethic (grounded in the self) for example, is historically associated with the rise of *laissez faire* capitalism and the mechanistic worldview discussed in the previous chapter, and is the ethic of mainstream industrial capitalism today. A homocentric ethic (grounded in the social good) underlies those ecological movements whose primary goal is social justice for all people, such as social ecologists, left Greens, social and socialist ecofeminists, many Second and Third World environmentalists, and the mainstream sustainable development movement. An ecocentric ethic (grounded in the cosmos, or whole ecosystem) guides the thinking of most deep ecologists, spiritual ecologists, Greens, cultural ecofeminists, organic farmers, bioregionalists, and most indigenous peoples' movements. The following discussion is not an exhaustive description of ethics. It does not discuss valuable insights into ethics developed by thinkers such as Aristotle, Aquinas, Kant, Eastern philosophers, or feminist philosophers (on the latter two see Chapters 4, 5, and 8). Rather it is an effort to develop some of the important ethical categories relevant to the environmental topics discussed in this book (See Table 3.1).

Environmental ethics are a link between theory and practice. They translate thought into action, worldviews into movements. Ideas generated from social conditions must be transformed into behaviors in order to change those conditions. Behaviors are thus guided by an underlying ethic. Religious beliefs, according to anthropologist Clifford Geertz, establish powerful moods and motivations that translate into social behaviors. Similarly, worldviews (whether mechanical or organic) asserts Charles Taylor, have powerful, associated sets of values that can override social changes and maintain existing social hegemony or be undermined, weakened, and transformed by social change and social movements. Ideas are thus translated into bodily motions that affect production and reproduction[2] (see Figure I.1).

EGOCENTRIC ETHICS

An egocentric ethic is grounded in the self. It is based on an individual ought focused on individual good. In its applied form, it involves the claim that what is good for the individual will benefit society. The individual good is thus prior to the social good which follows from it as a necessary consequence. An egocentric ethic's orientation does not derive from selfishness or narcissism, but rather is based on a philosophy that treats individuals (or private corporations) as separate, but equal, social atoms. Historically, the egocentric ethic rose to dominance in western culture during the seventeenth century. As the classic ethic of liberalism and *laissez faire* capitalism, in America it has been the guiding ethic of private entrepreneurs and corporations whose primary goal is the maximization of profit from the development of natural resources. Only the "silken bands of mild government", as Hector St John de Crèvecoeur put it in 1782, inhibit individual actions. Industry is "unfettered and unrestrained, because each person works for himself."[3]

Environmentally, an egocentric ethic permits individuals (or corporations) to extract and use natural resources to enhance their own lives and those of other members of society, limited only by the effects on their neighbors. Traditionally, the use of fire, common water sources, and rivers were regulated by laws. Under common law during the American colonial period, for example, one could not obstruct a river with a dam since this interfered with its natural course and reduced the privileges of others living along it. By the late eighteenth century, however, individual privileges increasingly prevailed when profits were at stake. Entrepreneurs could erect dams on the grounds that "the public whose advantage is always to be regarded, would be deprived of the benefit which always attends competition and rivalry."[4]

Egocentric ethics often reflect the Protestant ethic. An individual is responsible for his or her own salvation through good actions. During the seventeenth century, American Christianity moved away from the doctrine of the early Puritans that only the elect would be saved, toward the Arminian doctrine that any individual could assure

Table 3.1

Grounds for Environmental Ethics

Self: Egocentric		Society: Homocentric		Cosmos: Ecocentric	
Self-Interest	Religious	Utilitarian	Religious	Eco-Scientific	Eco-Religious
Thomas Hobbes	Judeo-Christian ethic	J.S. Mill	John Ray	Aldo Leopold	American Indian
John Locke	Arminian "heresy"	Jeremy Bentham	William Derham	Rachel Carson	Buddhism
Adam Smith		Gifford Pinchot	René Dubos	Deep ecologists	Spiritual Feminists
Thomas Malthus		Peter Singer	Robin Attfield	Restoration ecologists	Spiritual Greens
Garrett Hardin		Barry Commoner		Biological control	Process philosophers
		Murray Bookchin		Sustainable agriculture	
		Social ecofeminists			
		Left Greens			

Grounds for Obligation

Self-Interest	Religious	Utilitarian	Religious	Eco-Scientific	Eco-Religious
Maximization of individual self-interest: what is good for each individual will benefit society as a whole. Mutual coercion mutually agreed upon	Authority of God Genesis I Protestant ethic Individual salvation	Greatest good for the greatest number of people Social justice Duty to other humans	Stewardship by humans as God's caretakers Golden Rule Genesis II	Rational, scientific belief system based on laws of ecology Unity, stability, diversity, harmony of ecosystem Balance of nature or chaotic systems approach	Faith that all living and nonliving things have value Duty to whole environment Human and cosmic survival

(continued)

Table 3.1 (continued)

Grounds for Environmental Ethics

Self: Egocentric		Society: Homocentric		Cosmos: Ecocentric	
Self-Interest	Religious	Utilitarian	Religious	Eco-Scientific	Eco-Religious
		Metaphysics			
Mechanism		Both mechanistic and holistic		Organicism (Holism)	

Mechanism

1. Matter is composed of atomic parts
2. The whole is equal to the sum of the parts (law of identity)
3. Knowledge is context-independent
4. Change occurs by the rearrangement of parts
5. Dualism of mind and body, matter and spirit

Organicism (Holism)

1. Everything is connected to everything else
2. The whole is greater than the sum of the parts
3. Knowledge is context-dependent
4. The primacy of process over parts
5. The unity of humans and nonhuman nature

his or her own salvation through leading an ethical life.[5] In the seventeenth century, the Protestant ethic dovetailed with the Judeo–Christian mandate of Genesis I, 28: "Be fruitful and multiply, and replenish the earth and subdue it." From an environmental perspective, as historian Lynn White Jr argues, the Judeo–Christian ethic legitimated the domination of nature.[6] Early economic development in America was reinforced by this biblical framework. As the Arabella, bearing the first Puritan settlers of the Massachusetts Bay colony, left England for the New World in 1629, John Winthrop quoted the Genesis I passage.[7] In justifying American expansion into Oregon in 1846, John Quincy Adams asserted that the objectives of the United States were to "make the wilderness blossom as the rose; to establish laws, to increase, multiply, and subdue the earth, which we are commanded to do by the first behest of the God Almighty."[8] And Thomas Hart Benton that same year, in his famous address to the 29th Congress, insisted that the white race had "alone received the divine command to subdue and replenish the earth: for it is the only race that . . . hunts out new and distant lands, and even a New World, to subdue and replenish."[9] Similar Biblical passages reinforced God's command to transform nature from a wilderness into a civilization. Reverend Dr Dwinell's sermon, commemorating the joining of the Central Pacific and Union Pacific railroads in 1869, quoted the Bible as a sanction for human alteration of the natural landscape. "Prepare ye the way of the Lord, make straight in the desert a highway before our God. Every valley shall be exalted, and every mountain and hill shall be made low and the crooked shall be made straight and the rough places plain."[10]

Egocentric ethics as a basis for environmental policy are rooted in the philosophy of seventeenth century political philosopher Thomas Hobbes. In turn Hobbes' approach forms the ground for the environmental ethic of ecologist Garrett Hardin, whose "Tragedy of the Commons" (1968) influenced environmental policy in the 1970s.[11] For Hobbes, humans are basically competitive. In *Leviathan* (1651), Hobbes asserts that people are by nature unfriendly, hostile, and violent. In the state of nature, everyone has an equal right to everything, for "Nature has given all to all." But for Hobbes, nature is not a garden of Eden or a Utopia in which everyone shares its fruits, as earlier communal theories of society held. Instead, everyone is competing for the same natural resources. In *De Cive* (1647), he wrote, "For although any man might say of every thing, this is mine, yet he could not enjoy it, by reason of his neighbor, who having equal right and equal power, would pretend the same thing to be his."[12] Thus, because of competitive self-interest, the commons could not be shared, but must be fought over.

By Hobbes' time, the English commons were losing their traditional role as shared sources of life-giving grass, water, and wood to be used by all peasants, as had been the case in feudal Europe. Instead, they could be owned and enclosed by individual landlords who could use them to graze sheep for the expanding wool market. In fact, if lords did not compete, they could lose their lands and fortunes and be ridiculed by their peers. "For he that should be modest and tractable and perform all he promises," wrote Hobbes, ". . . should but make himself a prey to others and procure his own certain ruin."[13]

The commons was thus like a marketplace or a battleground in need of law and order. The solution to the disorder that prevailed in the state of nature was the social contract. By common consent, people gave up their freedom to fight and kill and out of fear accepted governance by a sovereign. Through the rational acceptance by each citizen of a set of rules for individual ethical conduct, social order, peace, and control could be maintained. The state was thus an artificial ordering of individual parts, a Leviathan, "to which we owe . . . our peace and defense."[14] Hobbes' egocentric ethic therefore was based on the assumption that human beings, as rational agents, could overcome their "natural" instincts to fight over property.

Garrett Hardin's "Tragedy of the Commons" and his "lifeboat ethics" are both grounded in this egocentric ethic. Like Hobbes, Hardin's (unstated) underlying assumptions are that people are naturally competitive, that capitalism is the "natural" form of economic life, and that the commons is like a marketplace. Hardin argues that individuals tended to graze more and more sheep on the commons because the economic gain was +1 for each sheep. On the other hand, the cost of overgrazing (environmental deterioration) was much less than −1, because the costs were shared equally by all. Thus there was no incentive to reduce herds. In the modern analogy, the seas and air are a global commons. Resource depletion and environmental pollution of the commons are shared by all, hence there is no incentive for individuals or nations to control their own exploitation. The costs of acid rain and chlorofluorocarbons in the air, oil spills and plastics in the oceans, and depletion of fish, whales, and seals are shared equally by all who fish, breathe, and live. The solution, for Hardin as for Hobbes, is mutual coercion, mutually agreed upon. People, corporations, and nation states voluntarily consent to rational regulation of resources.[15]

Similarly Hardin's "Living on a Lifeboat" (1974) is an egocentric ethic. When an overloaded boat capsizes, there will be insufficient lifeboats to save all. Those individuals who are saved are those who are strong enough to help themselves. When a population outstrips its food resources, some individual nations will institute population control policies and some will not. Through a policy of triage, such as that developed for wartime injury victims, selective help should be offered.[16] Under triage, limited wartime medical resources should be used first to help those with severe injuries who can survive only with aid and second to those with moderate injuries who would survive anyway. Those with massive fatal injuries who would die despite medical aid should not be helped beyond pain reduction. Similarly, developed nations with food surpluses should help developing nations which voluntarily agree to control population growth. Those who cannot or will not agree to population control policies should not receive assistance. The lifeboat ethic is thus an egocentric ethic of individual choice based on human reason. Nations, like individual atoms, are rational decision-makers who can decide whether or not to save themselves. Having arrived at that choice through reason, they voluntarily submit to coercion, i.e., population control, in order to save their countries.[17]

Egocentric ethics are rooted in the mechanistic science of the seventeenth century. Mechanism is based on several underlying assumptions consistent with liberal social theory:

1. Mechanistic science is based on the assumption that matter is made up of individual parts. Atoms are the real components of nature, just as individual humans are the real components of society.

2. The whole is equal to the sum of the individual parts. The law of identity in logic, or a = a, is the basis for the mathematical description of nature. Similarly, society is the sum of individual rational agents, as in Hobbes' depiction of the body of the sword-carrying sovereign as made up of the sum of the individual humans who have submitted themselves to his rule.

3. Mechanism involves the assumption of context independence. Real objects obey the laws of falling bodies and gravitation only when environmental contexts such as air resistance and friction are stripped away and masses act as point centers of force. In society, rules and laws are obeyed by a populace comprising equal individuals, stripped of particularity and difference.

4. Change occurs by the rearrangement of parts. In the billard ball universe of mechanistic scientists, the initial amount of motion (or energy) introduced into the universe by God at its creation is conserved and simply redistributed among the parts as they come together or separate to form the bodies of the phenomenal world. Similarly, individuals in society associate and dissociate in corporate bodies or business ventures.

5. Mechanistic science is often dualistic. Philosophers such as René Descartes and scientists such as Robert Boyle and Isaac Newton posited a world of spirit separate from that of matter. Nature, the human body, and animals could all be described, repaired, and controlled, as could the parts of a machine, by a separate human mind acting according to rational laws. Similarly, in the rhetoric of the founders of the American constitution, democratic society is a balance of powers as in a pendulum clock, and government operates as do the well-oiled wheels and gears of a machine controlled by human reason. Mind is separate from and superior to body; human society and culture are separate from and superior to nonhuman nature. Just as mechanistic science gives primacy to the individual parts that make up a corporeal body, so egocentric ethics give primacy to the individual humans that make up the social whole.

An egocentric ethic may be identified as the underlying ethic of private developers in current environmental disputes. Here the goals of entrepreneurs dedicated to promoting the individual's good conflict with those of government agencies charged with preserving the public good, and with those of environmentalists defending the good of nonhuman nature. Thus discharges of toxic chemicals by computer chip manufacturers in "Silicon Valley" on the San Francisco peninsula conflict with the regulatory mandates of water quality control agencies protecting groundwater quality. The efforts of Dow Chemical Corporation to locate a chemical processing plant in the Suisun Marsh area of the San Francisco Bay conflict with the public interest ethics of air and water quality control boards, and with the ecocentric ethics of environmentalists who wish to preserve the marsh as habitat for the endangered salt marsh harvest mouse.

From an environmental point of view, the egocentric ethic that legitimates *laissez faire* capitalism has a number of limitations. Because egocentric ethics are based on the assumption that the individual good is the highest good, the collective behavior of human groups or business corporations is not a legitimate subject of investigation. Second, because it includes the assumption that humans are "by nature" competitive and capitalism is the "natural" form of economics, ecological effects are external to human economics and cannot be adjudicated. In the nineteenth century, however, the first of these problems was dealt with through a new form of environmental ethics—the homocentric or utilitarian ethic. In the twentieth century, the problem of internalizing ecological externalities was addressed through the development of ecocentric ethics.

HOMOCENTRIC ETHICS

A homocentric (or anthropocentric) ethic is grounded in society. A homocentric ethic underlies the social interest model of politics and the approach of environmental regulatory agencies that protect human health. The utilitarian ethics of Jeremy Bentham (1789) and John Stuart Mill (1861), for example, advocate that a society ought to act in such a way as to insure the greatest good for the greatest number of people. The social good should be maximized, social evil minimized. For both Bentham and Mill, the utilitarian ethic has its origins in human sentience. Feelings of pleasure are good, those of pain are evil and to be avoided. Because people have the capacity for suffering, society has an obligation to reduce suffering through policies that maximize social justice for all.[18]

Utility, according to Bentham, "is that property in any object whereby it tends to produce benefit, advantage, good, or happiness . . . or to prevent the happening of mischief, pain, evil, or unhappiness." For Bentham the interest of the community is the "sum of the interests" of the individuals that compose it and actions are good in conformity with their tendency to "augment the happiness of the community." While Bentham spoke of the community and the sum of the individual interests that make up this "ficticious body," Mill cast his arguments in terms of the "general interests of society," "the interest of the whole," and "the good of the whole."[19] Each individual, he assumed, is endowed with feelings that promote the general good. "Utilitarian morality recognizes in humans the power of sacrificing their own greatest good for the good of others." Each person should associate his or her happiness with "the good of the whole." People therefore have primary duties and obligations to other humans, not just to themselves.[20] "Actions," he said, "are right in proportion to as they tend to promote happiness; wrong as they tend to produce the reverse of happiness."[21]

In developing an ultimate sanction for the principle of utility, Mill went beyond the simple prohibitions against killing and robbery in the Mosaic decalogue and the Hobbesian idea that it is "natural" for individuals freely to kill each other unless they give up that right and receive protection from a sovereign. "I feel I am bound not to rob or murder, betray or deceive; but why am I bound to promote the general happiness?" he asked. The answer lies in education. The more "education and general cultivation," the more powerful is the enforcement. Education overcomes selfish motives and creates deeply rooted feelings of unity with other humans. Moral feelings are not innate, but acquired. Mill claimed that a sequence of ethical standards develops as "civilization" advances and mankind is "further removed from a state of savage independence." The spirit of the utilitarian ethic is expressed in the Golden Rule. " 'To do as you would be done by,' and 'To love your neighbor as yourself,'" Mill wrote, "constitute the ideal perfection of utilitarian morality."[22]

As in egocentric ethics, this homocentric ought reflects a religious formulation. Humans are stewards and caretakers of the natural world. Scholars such as ecologist René Dubos and philosophers John Passmore and Robin Attfield have pointed out that the Bible contains numerous passages that countervene the stark domination ethic of Genesis 1.[23] In Genesis 2, thought to be derived from a different historical tradition than Genesis 1, the animals are helpmeets for humans. God, according to Dubos, "placed man in the Garden of Eden not as a master but rather in a spirit of stewardship."[24] Like egocentric ethics, stewardship ethics were enunciated by seventeenth century scientists and theologians concerned about the atheistic implications of mechanism as formulated by Hobbes. John Ray and William Derham developed a theology of stewardship consistent with Newtonian science, human progress, and the management of nature for human benefit. They quote New Testament passages, such as Matthew (25:14): "That these things are the gifts of God, they are so many talents entrusted with us by the infinite Lord of the world, a stewardship, a trust reposed in us; for which we must give an account at the day when our Lord shall call." Additionally, in Luke 16:2, God said to the unfaithful steward, "Give an account of thy stewardship, for thou mayest no longer be steward." In stewardship ethics, God as the wise conservator and superintendent of the natural world made humans caretakers and stewards in his image. Stewardship ethics, however, are fundamentally homocentric. Humans must manage nature for the benefit of the human species, not for the intrinsic benefit of other species.[25]

Like egocentric ethics, homocentric ethics are consistent with the assumptions of mechanistic science, especially as extended by nineteenth century scientists to include the fields of thermodynamics, hydrology, and electricity and magnetism. Scientific experts could use these laws for the efficient management of natural resources. Yet certain assumptions that characterize later ecocentric ethics are melded with the homocentric. Both nature (as in Darwinian evolution) and society are described in terms of organic metaphors. As Supreme Court Justice Oliver Wendell Holmes Jr put it in 1903, "In modern societies, every part is so organically related to every other part, that what affects any portion must be felt more or less by all the rest."[26]

In addition to the utilitarian philosophers, Marxists espouse a homocentric ethic. Inasmuch as Marx's goals were to better the human condition by using science and technology to meet human needs for food, clothing, shelter, and fuel and to overcome the necessities imposed by nature, his philosophy is clearly human centered. Based on Marx's ideas and critique as fully cognizant of the disruptions of nonhuman nature by capitalist industries, post-Marxist social ecologists, such as Barry Commoner and Murray Bookchin, present political alternatives that mitigate problems of resource depletion and pollution. They see scientific research as developing out of capitalist social hierarchies and industrial and university relations. They offer technologies and social structures designed to keep human needs in balance with natural cycles and with energy requirements. A homocentric ethic guides choices concerning which research projects to fund, which technologies to implement, and which processes to use for decision-making. Such an ethic sets up the fulfillment of human needs as a priority, but gives full consideration to nonhuman nature in the process of decision making. Homocentric ethics underlie the politics of the social ecologists of Chapter 6.

What are some examples of homocentric ethics and political conflict? A particularly salient example is the building of dams for water and hydraulic power for cities and states. The controversy in the early twentieth century over whether to dam Hetch Hetchy valley in Yosemite Park as a source for water and power for the city of San Francisco was won by utilitarians. Gifford Pinchot, arguing for San Francisco, pointed out that a water supply for the city was a greater good for a greater number of people than leaving the valley in the state of nature for a few hikers and nature lovers. John Muir, on the other hand, viewed the valley as one of God's cathedrals and the proponents of the dam as temple destroyers, an ethic based on the valley's intrinsic right to remain as created. Today water control agencies are quite explicit in their claim that they must consider the greatest good for the greatest number of people in distributing water to their customers in time of shortages.[27]

In 1979 environmentalist Mark Dubois chained himself to a rock to prevent California's Stanislaus River from being dammed and losing its right to remain free. "All the life of this canyon, its wealth of archaeological and historical roots to our past, and its unique geological grandeur are enough reasons to protect this canyon just for itself," he wrote to the Army Corps of Engineers. "But in addition, all the spiritual values with which this canyon has filled tens of thousands of folks should prohibit us from committing the unconscionable act of wiping this place off the face of the earth." This controversy may be viewed as a conflict among interest groups with different underlying ethics. Farmers and corporate agribusiness ventures, whose egocentric ethics promote the individual's good, along with federal water control agencies, whose homocentric ethics see water development as the greatest good for the greatest number, conflict with the ecocentric ethics of environmentalists, who support the river's intrinsic right to remain wild.[28]

Dilemmas such as these point up one of the main problems of both egocentric and homocentric ethics—their failure to internalize ecological externalities. Ecological changes and their long-term effects are outside the human/society framework of these ethics. The effects of ecological changes such as salinity build-up in farming soils that use the dam's water, or the loss of indigenous species when a valley is flooded, are not part of the human-centered calculus of decision making. One approach offered by ethicists is to extend homocentric ethics to include other sentient species. Animal liberationists Peter Singer and Tom Regan, for example, extend the pleasure-pain principle of Bentham and Mill to animals, arguing that conditions for the well-being of animals should be maximized, while conditions that lead to pain such as over-crowded conditions, liquid diets, and cruel experimentation should be minimized.[29] A similar extension of stewardship ethics to include nonhuman species and future human beings is made by Robin Attfield.[30] An alternative, however, is to formulate a radically different form of environmental ethics—ecocentric ethics.

ECOCENTRIC ETHICS

An ecocentric ethic is grounded in the cosmos. The whole environment, including inanimate elements, rocks, and minerals along with animate plants and animals, is assigned intrinsic value. The eco-scientific form of this ethic draws its ought from the science of ecology. Recognizing that science can no longer be considered value-free, as the logical positivists of the early-twentieth century had insisted, proponents of ecocentric ethics look to ecology for guidelines on how to resolve ethical dilemmas. Maintenance of the balance of nature and retention of the unity, stability, diversity, and harmony of the ecosystem are its overarching goals. Of primary importance is the survival of all living and non-living things as components of healthy ecosystems. All things in the cosmos as well as humans have moral considerability.

Modern ecocentric ethics were first formulated by Aldo Leopold during the 1930s and 1940s and published as "The Land Ethic," the final chapter of his posthumous *A Sand County Almanac* (1949). Some of Leopold's inspiration for the land ethic seems to have derived from Mill's *Utilitarianism*. Like Mill—who wrote about the "influences of advancing civilization," the "removal from the state of savage independence," and the utilitarian Golden Rule as superseding the basic prohibitions against robbing and murdering—Leopold thought ethics developed in sequence. "The first ethics," he wrote, "dealt with the relation between individuals; the Mosaic Decalogue is an example. Later accretions dealt with the relation between the individual and society. The Golden Rule tries to integrate the individual to society." The land ethic, he argued, extends the sequence a step further. It enlarges the bounds of the community to include "soils, waters, plants, and animals, or collectively, the land." It "changes the role of *homo sapiens* from conqueror of the land-community to plain member and citizen of it. It implies respect for his fellow members and also respect for the community itself."[31] In putting the land ethic into practice, Leopold urged that each question be judged according to what is both ethically and aesthetically right. Perhaps influenced by Mill's phraseology that "actions are right in proportion as they tend to promote happiness; wrong as they tend to produce the reverse of happiness," Leopold wrote: "A thing is right when it tends to preserve the integrity, beauty, and stability of the biotic community. It is wrong when it tends otherwise." Like Mill who argued for the importance of education in creating obligations toward other people, Leopold argued that in order to overcome economic self-interest, ethical obligations toward the land must by taught through conservation education.[32]

Environmental historian Roderick Nash has elaborated Leopold's land ethic in an article "Do Rocks have Rights?" Rocks are part of the pyramid of animate and inanimate things governed by the laws of ecology. Even though rocks are not sentient like animals, rocks as well as plants can be assigned interests that can be represented and adjudicated. Yet such a concept might still be used to protect rocks in the interest of humans. Pushing it further, Nash argues, we can "suppose that rocks, just like people, do have rights in and of themselves. It follows that it is the rock's interest, not the human interested in the rock, that is being protected." Other cultures such as Native Americans, Zen Buddhists, and Shintos, he points out, assume that rocks are alive—a mystical religious belief not usually held by western philosophers and scientists.[33]

Ecocentric ethics are rooted in a holistic, rather than mechanistic, metaphysics.[34] The assumptions of holism are:

1. Everything is connected to everything else. The whole qualifies each part; conversely, a change in one of the parts will change the other parts and the whole. Ecologically, this has been illustrated by the idea that no part of an ecosystem can be removed without altering the dynamics of the cycle. If too many changes occur, an ecosystem collapses. Alternatively, to remove the parts from the environment for study in the laboratory may result in a distorted understanding of the ecological system as a whole.[35]

2. The whole is greater than the sum of the parts. Unlike the concept of identity in which the whole equals the sum of the parts, ecological systems experience synergy: the combined action of separate parts may produce an effect greater than the sum of the individual effects. This can be exemplified by the dumping of organic sewage and industrial pollutants into lakes and rivers. The bacterial increases may cause those drinking or swimming in the water to become ill. But if the bottom of the lake is covered with metallic mercury, the overall hazard is more than doubled because the bacteria may also transform the metallic mercury into toxic methyl mercury which becomes concentrated in the food chain.[36]

3. Knowledge is context-dependent. As opposed to the context independence assumption of mechanism, in holism each part at any instant takes its meaning from the whole. For example, in a hologram, produced by directing laser light through a half-silvered mirror, each part of the three-dimensional image contains information about the whole object. There are many-to-one and one-to-many relationships, rather than the point to point correspondences between object and image found in classical optics. Similarly, in perception, objects are integrated patterns. The whole is perceived first with an awareness of hidden aspects, background, and recognition of patterns, as when one views a tree or a house.[37]

4. The primacy of process over parts. As opposed to the closed, isolated equilibrium and near-equilibrium systems studied in classical physics (such as the steam engine), biological and social systems are open. These are steady-state systems in which matter and energy are constantly being exchanged with the surroundings. Living things are dissipative structures, resulting from a continual flow of energy, just as a vortex in a stream is a structure arising from the continually changing water molecules swirling through it. Ilya Prigogene describes an open, far-from-equilibrium thermodynamics in which new order and organization can arise spontaneously. Nonlinear relationships occur in which small inputs can spontaneously produce large effects (see Chapter 4).[38]

Continual change and process are not only significant in ecology, but also are fundamental to the new physics. Physicist David Bohm in his book *Wholeness and the Implicate Order* (1980) describes process as originating from an undivided multidimensional wholeness called a holomovement (see Chapter 4). Within the holomovement is an implicate order that unfolds to become the explicate order of stable, recurring elements observed in the everyday world. The holomovement is life-implicit, the ground of both inanimate matter and of life.[39]

5. The unity of humans and nonhuman nature. As opposed to nature/culture dualism, in holism humans and nature are part of the same organic cosmological system. While theoretical ecologists often focus their research on natural areas removed from human impact, human (or political) ecologists study the mutual interactions between society and non-human nature.

Just as mechanism dovetailed with certain political assumptions, so holism has been seen to imply particular kinds of politics. Holism found favor among philosophers and ecologists during the 1920s. In the 1930s, however, its emphasis on the whole over and above the parts was viewed as being consistent with fascism. This contributed to the replacement of holistic and organismic assumptions in biology by mechanistic modes of description. In the 1960s and 1970s holistic ideas returned, with the blossoming of small-scale back-to-the-land communes and households in which decision-making was vested in the consensus of the whole group. Recently the emergence of green politics has given rise to a political movement dedicated to the establishment of an ecologically viable society (see Chapter 7). Drawing on holistic assumptions, the bioregional movement emphasizes living within the resources of the local watershed and developing them to sustain the human and nonhuman community as an ecological whole (see Chapter 9). Ecocentric ethics also have religious and spiritual components. Deep ecology, nature religions, ecological spirituality, and process philosophy have at their roots an ecocentric value system (see Chapters 4 and 5).[40]

Ecocentric ethics, like egocentric and homocentric ethics, have a number of philosophical difficulties. Finding a philosophically adequate justification for the intrinsic value of non-human beings has been called by some environmental philosophers the central axiological problem of environmental ethics. In mainstream Western culture, only human beings have traditionally had inherent worth, while the rest of nature has been assigned instrumental value as a resource for humans. Thus, within an egocentric or homocentric ethic, it is not *morally* wrong to kill or use the last of a species of animal, plant, or mineral when human survival is at stake. Within an ecological ethic, however, such a decision could depend on finding an adequate justification for the intrinsic value of the nonhuman species, as well as on the particular circumstances. At bottom, ecocentric ethics may have a homocentric justification.[41]

A second problem stems from the distinction between facts and values. The separation of observable facts from humanly assigned values, or *is* from *ought*, has been a mainstay of western science since the work of David Hume in the eighteenth century. Can a property such as the goodness or richness of animals, rocks, or the biosphere be inferred through the senses as an objective, intrinsic characteristic of the entities in question? Can there properly be such a such a thing as an ecological ethic, when ecology is an objective science and ethics is a subjective value system?

Environmental philosophers have proposed a number of answers to these questions. One approach is to question the possibility that facts can be separated from values in science and philosophy. Another is to recognize that descriptions of what *is* can include intrinsic value, while questions of what one ought to do belong to a different category.

Proponents of intrinsic value in nature include Holmes Rolston, III and J. Baird Callicott. Rolston argues that intrinsic values are objective and actually found in nature. Yet the connection between nature and values is complex. Scientific descriptions of nature and values arise together:

> Ecological description finds unity, harmony, interdependence, stability, etc. and these are valuationally endorsed. . . . In post-Darwinian nature . . . we looked for these values in vain, while with ecological description we now find them . . . here an "ought" is not so much *derived* from an "is" as discovered with it. As we progress from descriptions of fauna and flora, of cycles and pyramids, of stability and dynamism, on to intricacy, planetary opulence and interdependence, to unity and harmony . . . arriving at last at beauty and goodness, it is difficult to say where the natural facts leave off and where the natural values appear . . . The sharp is/ought dichotomy is gone; the values seem to be there as soon as the facts are fully in, and both alike are properties of the system.[42]

For Rolston, science is objective truth whose sphere continually expands with greater human knowledge. Darwinian evolution is not different from, but encompassed by, an emerging science of ecology. Yet values are inherent in science itself and not separate from it. They are discovered in nature simultaneously with objective truths about it. Rolston's approach is an inversion of a social constructivist viewpoint. The social construction of science would argue that both truths and values are deeply intertwined, but they are imposed on nature by humans inbedded in a value system derived from their place in class society and in social history.

Callicott's primary concern is to derive a philosophical basis for assigning intrinsic rather than merely instrumental value to nonhuman nature. If nonhuman species can be shown to have intrinsic value, then it follows as a powerful corollary that other species have a right to exist in and of themselves and that humans have moral obligations to them. In searching for a basis for intrinsic value in nature, Callicott offers the argument that quantum mechanics could provide such a foundation. If quantum theory forces us to abandon the sharp dichotomy between subject and object characteristic of the Cartesian/ Newtonian mechanistic worldview, then David Hume's sharp distinction between valuing subjects and value-free objects, or between is and ought must also be abandoned. Thus from revolutionary changes in science follow revolutionary changes in ethics. Such considerations also lead to a breakdown of the dualism between self and world. "Since nature is the self fully extended and diffused, and the self, complementarily, is nature concentrated and focused in one of the intersections, the 'knots,' of the web of life, . . . nature is intrinsically valuable, to the extent that the self is intrinsically valuable."[43]

Like Rolston, Callicott accepts the evolution of science as providing ever greater access to truth. We simply expand our understanding of the fact that nature now includes both subject and object intertwined and interacting. We evolve from a mechanistic framework for grounding ethics to a quantum mechanical framework. But neither philosopher takes into consideration the social or political basis on which much of science is constructed and in which its theories are imbedded.

Yet another difficulty with Aldo Leopold's and Roderick Nash's formulation of ecocentric ethics lies in their supposition of the development of sequential ethics. The advancement of civilization does not necessarily imply the evolution of more sophisticated ethics. The assumption that the earliest ethics dealt with the relations between individuals imposes the assumptions of Hobbes' hypothetical "state of nature" and the individualism of *laissez faire* capitalism onto the earliest peoples.[44]

Finally, some feminists and persons of color criticize ecocentrism for a holism that masks a gender, racial, and species difference. Ecocentric ethics make each individual—whether mosquito or person, male or female, white or black—equally subordinate to the over-arching whole.[45]

CONCLUSION

Despite these underlying difficulties, egocentric, homocentric, and ecocentric environmental ethics have all received attention and development since the environmental movement of the 1970s and 1980s. In conflicts of interest over environmental and quality of life issues, these catagories are useful in analyzing the implicit ethical positions assumed by both mainstream and radical political groups. Variations of these three forms of environmental ethics underlie the political positions of the environmental theorists of Part II and the environmental activists of Part III. Environmental ethics link the ideas of the theorists with the movements of the activists, translating ideas into behaviors in the effort to bring about a livable world.

FURTHER READING

Attfield, Robin. *The Ethics of Environmental Concern.* New York: Columbia University Press, 1983.

Bentham, Jeremy. *An Introduction to the Principles of Morals and Legislation* [1789]. London: W. Pickering, 1823.

Callicott, J. Baird, ed. *Companion to A Sand County Almanac: Interpretative and Critical Essays.* Madison: University of Wisconsin Press, 1987.

———. *In Defense of the Land Ethic: Essays in Environmental Philsophy.* Albany: State University of New York Press, 1989.

Flader, Susan. *Thinking Like a Mountain: Aldo Leopold and the Evolution of an Ecological Attitude toward Deer, Wolves, and Forests.* Lincoln: University of Nebraska Press, 1974.

Hardin, Garrett, and John Baden, eds. *Managing the Commons.* San Francisco: W. H. Freeman, 1977.

———. *Promethean Ethics: Living with Death, Competition, and Triage.* Seattle: University of Washington Press, 1980.

Hargrove, Eugene. *Foundations of Environmental Ethics.* Englewood Cliffs, N. J.: Prentice Hall, 1989.

Leopold, Aldo. *A Sand County Almanac.* London: Oxford University Press, 1949.

Leopold, Aldo. *The River of the Mother of God and Other Essays.* Edited by Susan Flader and Baird Callicott. Madison, Wi.: University of Wisconsin Press, 1991.

Meine, Curt. *Aldo Leopold: The Man and His Work*. Madison, Wi.: University of Wisconsin Press, 1989.

Mill, John Stuart. *Utilitarianism* [1861]. Indianapolis: Bobbs Merrill, 1957.

Nash, Roderick. *The Rights of Nature*. Madison: University of Wisconsin Press, 1989.

Passmore, John. *Man's Responsibility for Nature*. New York: Scribner's, 1974.

Phillips, D. C. *Holistic Thought in Social Science*. Stanford, Ca.: Stanford University Press, 1976.

Regan, Tom. *All That Dwell Therein—Essays on Animal Rights and Environmental Ethics*. Berkeley, Ca.: University of California Press, 1982.

————. *The Case for Animal Rights*. Berkeley, Ca.: University of California Press, 1983.

Rolston III, Holmes. *Environmental Ethics: Duties to and Values in the Natural World*. Philadelphia, Pa.: Temple University Press, 1988.

————. *Philosophy Gone Wild: Essays in Environmental Ethics*. Buffalo, N. Y.: Prometheus Books, 1986.

Singer, Peter. *Animal Liberation: A New Ethics for our Treatment of Animals*. New York: Avon, 1975.

Stone, Christopher. *Earth and Other Ethics: The Case for Moral Pluralism*. New York: Harper and Row, 1987.

————. *Should Trees Have Standing: Toward Legal rights For Natural Objects*. Los Altos, Ca.: William Kaufmann, 1974.

Taylor, Paul. *Respect for Nature: A Theory of Environmental Ethics*. Princeton, N.J.: Princeton University Press, 1986.

II

THOUGHT

II

THOUGHT

4

DEEP ECOLOGY

Deep ecologists call for a new ecological paradigm that will replace the dominant mechanistic paradigm of the past three hundred years. This new worldview would represent as profound a transformation as the one which occurred during the scientific revolution of the seventeenth century. It would be so fundamental that it would entail new metaphysical, epistemological, religious, psychological, sociopolitical, and ethical principles. Taking its name and approach from Norwegian philosopher Arne Naess' 1972 article on "The Shallow and the Deep, Long-Range Ecology Movement," deep ecology holds that the reform environmentalism of the 1970s and 1980s dealt only with legal and institutional fixes for pollution and resource depletion, rather than fundamental changes in human relations with nonhuman nature (see Table 4.1).[1] When in 1985 sociologist Bill Devall and philosopher George Sessions published a book and writer Michael Tobias a collection of articles entitled *Deep Ecology,* the concept gained visibility beyond the community of philosophers. It has now become the legitimating framework for an array of ecological movements from spiritual Greens to radical Earth First!ers.

A dominant social paradigm, according to Bill Devall, who elaborated on Naess' approach, is a "mental image of social reality that guides expectations in a society." Deep ecology challenges the dominant western paradigm elaborated in Chapter 2. It offers a new science of nature, a new spiritual paradigm, and a new ecological ethic. Deep ecological thinking emerges from the sense of ecological crisis detailed in Chapter 1. It is thus socially produced and socially constructed. It focuses, however, on transformation at the level of consciousness and worldview, rather than the transformation of production and reproduction (see Figure I.1). It thus supports and legitimates new social and economic directions that move the world toward sustainability.

PRINCIPLES OF DEEP ECOLOGY

For Devall and Sessions, deep ecology requires a new metaphysics of humans-in-nature not above it. This cosmic/ecological metaphysics stresses an I/thou relationship between humans and nonhuman nature and the integrity of person/planet. The principle of biospheric equality places humans on an equal level with all other living things in an organismic democracy. Here it draws from the science of ecology which attributes equal importance to every component of the interlinked web of nature.

Second, a new psychology, or philosophy of self, is required. This means a total intermingling of person with planet. A society based on the prominence of individual egos gives way to a new spiritual freedom to develop an interconnected community. Urban intellects previously

Table 4.1

Arne Naess' Principles of Deep Ecology

1. Rejection of the man-in-environment image in favor of the relational, total-field image.
2. Biospherical egalitarianism.
3. Principles of diversity and of symbiosis.
4. Anti-class posture.
5. Fight against pollution and resource depletion.
6. Complexity, not complication.
7. Local autonomy and decentralization.

Source: Arne Naess, "The Shallow and the Deep, Long-Range Ecology Movement. A Summary," *Inquiry*, 16 (1972): 95–100.

dedicated to the self-consciousness of power over planet open them-selves to a person in planet consciousness. This avenue draws them down a Buddhist or Hinduist pathless path by which self can be integrated into the Great Self. Modesty and humility and an awe of evolution take precedence over an assertion of human power over the biosphere. Spiritual ecology and the spiritual wing of the Greens movement in the United States further develop this assumption (see Chapter 5).

Third, deep ecology develops a new anthropology that draws its guidelines from studies of horticulturalists and gatherer-hunters. Reinhabiting the land as "dwellers in it" rejects industrial society as the world paradigm for development and entails leaving vast tracts of land as wilderness. People can live their lives as "future primi-tive" withdrawing from developed land and allowing it to reestab-lish itself as wilderness. For each ecological region, the guideline for use should be human carrying capacity. Much of the thought underlying the bioregional movement stems from this assumption (see Chapter 9).

Fourth, deep ecology espouses an ecocentric rather than a homo-centric (or anthropocentric) ethic. In using nonhuman nature, people have a duty to maintain the integrity of the ecosphere, not to conquer it or make it more efficient. Although living entails some killing, other organisms have a right to exist and evolve just as do humans. Humans are dependent on the ecosphere for survival and should not exploit it as a master does a slave. This assumption is fundamental to an emerging ecocentric ethic rooted in ecologist Aldo Leopold's 1949 "land ethic."

Fifth, a new ecologically-based science promotes a sense of human place within the household of nature. A nonviolent peace with nature is declared. The new scientist takes her cue from the ancient shaman rather than the genetic engineer. The new science is process oriented. It draws on design with nature, rather than the imposition of form on nature. Biological and cultural diversity are desired ends. These can be reached and maintained through soft energy and appropriate technology paths. Technology is not an end but a means to human welfare.

Deep ecology's sources include alternative traditions in western thought as well as the beliefs of native peoples and eastern philosophers. In the Western religious tradition it espouses the teachings of Saint Francis of Assisi, rather than the Judeo-Christian tradition of domination over nature. From eastern philosophy it learns from interpreters and poets such as Alan Watts, Daisetz Suzuki, and Gary Snyder and draws on historian Joseph Needham's work on science and civilization in China. From Native American leaders such as Black Elk and Luther Standing Bear, it seeks a new religious ecology and social organization. Alternative western philosophers provide guidelines to the possibility of integrating humans within nature. These include the Presocratics, Giordano Bruno, Baruch Spinoza, Gottfried Wilhelm Leibniz, Henry David Thoreau, John Muir, George Santayana, Alfred North Whitehead, Aldo Leopold, Robinson Jeffers, and Martin Heidegger. Deep ecology draws its scientific inspiration from Paul Shepard's view that ecology is a subversive science—the basis of a social and scientific resistance movement.[2]

Another call for a "New Ecological Paradigm" (NEP) comes from sociologists William Catton and Riley Dunlap. Following Columbus' discovery of the New World, they argue, Europeans expanded "exuberantly" across America, to use the language of great plains historian Walter Prescott Webb, where the person/land ratio was ten times less than in Europe. An age of abundance and industrialization followed in which nature was exploited by a "people of plenty," who clung to an ideology of progress. The "Dominant Western Worldview" (DWW), which guided American development, assumed that people were different from all other organisms and were in charge of their own destiny. Because global resources were so abundant, and people had a unique capacity to develop and solve problems using technology, they believed they would always be able to find solutions that would continue humanity's forward progress. A corollary to this worldview, the "Human Exemptionalism Paradigm (HEP), assumed that human societies were exempt from the consequences of ecological principles and environmental constraints (see Table 4.2).

The ecological crisis and the growing awareness of resource scarcity, however, challenge these older assumptions. Catton and Dunlap suggest that a New Ecological Paradigm will replace the Dominant Western Worldview and the concept of human exemptionalism, ushering in a "post-exuberant age" (Figure 4.1). The ecological paradigm rests on an historically new set of assumptions about people and nature. The NEP assumes that although humans have unique characteristics as a species, they are still subject to the same ecological laws and restraints as other organisms. Humans are dependent on finite natural resources and there are important linkages and feedbacks between human societies and the ecosystems in which they are imbedded. If human technological progress exceeds the carrying capacity of the land, the laws of ecology will force adjustments. A steady state or sustainable society is "one that provides for successful human adaptation to a finite (and vulnerable) ecosystem on a long-term basis."[3]

A third push to establish a deep ecological paradigm comes from physicist Fritjof Capra. Famous both for his analysis of the similarity between the assumptions underlying the new physics and eastern philosophy (in *The Tao of Physics*) and his call for a revolution in thought patterns (in *The Turning Point*), he has embraced deep ecology as the most succinct term for the emerging worldview. The worldview that has dominated western society for the past three hundred years, he argues, assumes that the universe is made up of elementary particles, the human body is a machine, society is based on a Darwinian competitive struggle for existence, a belief in material progress, and that the female is subordinate to the male.[4]

Deep ecology, Capra believes, offers a holistic worldview that

Table 4.2

Environmental Paradigms

	Dominant Western Worldview (DWW)		Human Exemptionalism Paradigm (HEP)		New Ecological Paradigm (NEP)
Assumptions about the nature of human beings:	DWW$_1$	People are fundamentally different from all other creatures on Earth, over which they have dominion.	HEP$_1$	Humans have a cultural heritage in addition to (and distinct from) their genetic inheritance, and thus are quite unlike all other animal species.	NEP$_1$ While humans have exceptional characteristics (culture, technology, etc.), they remain one among many species that are interdependently involved in the global ecosystem.
Assumptions about social causation:	DWW$_2$	People are master of their destiny; they can choose their goals and learn to do whatever is necessary to achieve them.	HEP$_2$	Social and cultural factors (including technology) are the major determinants of human affairs.	NEP$_2$ Human affairs are influenced not only by social and cultural factors, but also by intricate linkages of cause, effect, and feedback in the web of nature; thus purposive human actions have many unintended consequences.
Assumptions about the context of human society:	DWW$_3$	The world is vast, and thus provides unlimited opportunities for humans.	HEP$_3$	Social and cultural environments are the crucial context for human affairs, and the biophysical environment is largely irrelevant.	NEP$_3$ Humans live in and are dependent upon a finite biophysical environment which imposes potent physical and biological restraints on human affairs.

(continued)

Table 4.2 (continued)

Environmental Paradigms

	Dominant Western Worldview (DWW)	Human Exemptionalism Paradigm (HEP)	New Ecological Paradigm (NEP)
Assumptions about constraints on human society:	DWW₄ The history of humanity is one of progress; for every problem there is a solution, and thus progress need never cease.	HEP₄ Culture is cumulative; thus technological and social progress can continue indefinitely, making all social problems ultimately soluble.	NEP₄ Although the inventiveness of humans and the powe's derived therefrom may seem for a while to extend carrying capacity limits, ecological laws cannot be repealed.

Source: William R. Catton, Jr. and Riley Dunlap. "A New Ecological Paradigm for Post-Exuberant Sociology," *American Behavioral Scientist*, 24, no. 1 (Sept/Oct. 1980): 34, reprinted by permission.

FIGURE 4.1
EXPONENTIAL AND LOGISTIC GROWTH MODELS

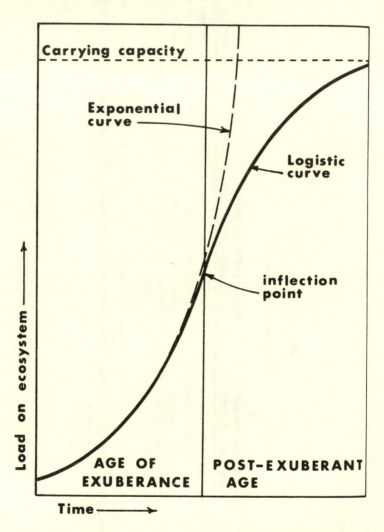

Source: William R. Catton, Jr. and Riley Dunlap. "A New Ecological Paradigm for Post-Exuberant Sociology," *American Behavioral Scientist*, 24, no. 1 (Sept/Oct 1980): 28, reprinted by permission.

emphasizes the whole over the parts and does not separate humans from the environment. The ecological paradigm entails a new ethic that recognizes the intrinsic value of all beings, one that will replace the anthropocentric ethics of the past. "All natural systems are wholes whose specific structures arise from the interactions and interdependence of their parts. Systemic properties are destroyed when a system is dissected, either physically or theoretically, into isolated elements. Although we can discern individual parts in any system, the nature of the whole is always different from the mere sum of its parts." Similarly, a new green economics sees the economy as a living system made up of interacting human beings and social organizations. Its goals are to maximize human health, welfare, basic needs, and the environment, rather than profit. A number of new social movements have embraced these goals, including the ecology, feminist, holistic health, human potential, and green movements.

The transition to a new worldview, Capra believes, coincides with a transformation in values that could bring about a balance between the rational and the intuitive, the reductionist and holistic, and the analytic and synthetic. The purpose is not to abandon one mode for the other, but to work toward a balance between them.[5]

SCIENTIFIC ROOTS OF DEEP ECOLOGY

Emerging over the past decade are a number of scientific proposals that challenge the scientific revolution's mechanistic view of nature. According to physicist David Bohm, a mechanistic science based on the assumption that matter is divisible into parts (such as atoms, electrons, or quarks) moved by external forces may be giving way to a new science based on the primacy of process. In the early twentieth century, he argues, relativity and quantum theory began to challenge mechanism. Relativity theory postulated that fields with varying strengths spread out in space. Strong, stable areas, much like whirlpools in a flowing stream, represented particles. They interacted with and modified each other, but were still considered external to and separate from each other. Quantum mechanics mounted a greater challenge. Motion was not continuous, as in mechanistic science, but occurred in leaps. Particles, such as electrons, behaved like waves, while waves, such as light waves, behaved like particles, depending on the experimental context. Context dependence, which was antithetical to mechanism and part of the organic worldview, was a fundamental characteristic of matter.

Bohm's process physics challenges mechanism still further. He argues that instead of starting with parts as primary and building up wholes as secondary phenomena, a physics is needed that starts with undivided, multidimensional wholeness (a flow of energy called the holomovement) and derives the three dimensional world of classical mechanics as a secondary phenomenon. The explicate order of the Newtonian world in which we live unfolds from the implicate order contained in the underlying flow of energy.

Bohm suggests that the holomovement contains the principle or seed of life that directs the environment as well as the energy that comes from the soil, water, air, and sunlight. Just as a forest contains trees that are continually being replaced by new ones, so a particle is in a stable, but continual state of regular changes that manifest over and over again. Living and inanimate things are similar in that they reproduce themselves over and over by unfoldment and enfoldment. When inanimate matter is informed by a seed containing information in its DNA, it produces a living plant which in turn reproduces a seed. The plant exchanges matter and energy with its environment; carbon dioxide and oxygen cross the cell boundaries. At no point is there a sharp distinction between life and nonlife. "The holomovement which is 'life-implicit,' says Bohm, "is the ground both of 'life-explicit' and of 'inanimate matter'. . . . Thus we do not fragment life and inanimate matter, nor do we try to reduce the former completely to nothing but an outcome of the latter."[6]

Another challenge to mechanism comes from the new thermodynamics of Ilya Prigogine. The clock-like machine model of nature and society that dominated the past three centuries of western thought may be winding down. While Newtonian classical physics is still valid, it is nonetheless limited to a clearly defined domain of the total world. It was extended in the nineteenth century to include theories of thermodynamics that developed out of the needs of a steam-engine society, electricity and magnetism that supplied the light and electricity that powered that society, and hydrodynamics or the science associated with the dams and water power that generated its electricity. The equilibrium and near-equilibrium thermodynamics of nineteenth-century classical physics had beautifully described closed, isolated sytems such as steam engines and refrigerators.

In dealing with the emergence of order out of chaos, Prigogine's theory helped to clarify an apparent contradiction between two nineteenth century scientific developments. Classical thermodynamics, which says that the universe is moving toward a greater state of chaos, is based on two laws. The first law states that the total energy of the universe is constant and only changes its form as it is transferred from mechanical, to chemical, to hydrodynamic, to metabolic energy etc. But the second law states that the energy available for work—the useful energy—is decreasing. The universe is running down, just as a clock unwinds over time when no one is there to rewind it. The second law implies that the world proceeds from order to disorder, that people grow older, and that in billions of years the whole universe will reach a uniform temperature. The classical model of reality deals very adequately with closed systems that are isolated from their environments—situations in which small inputs result in small outputs that can be described by linear mathematical relationships.

Yet the very concept of an unwinding clocklike universe is apparently contradicted by another startling nineteenth century theory—evolution, or the motion toward greater order. Darwinian evolution says that biological systems are evolving, not running down. They are moving from disorder to order; they are becoming more organized rather than disorganized. The direction of change over time is from simple to more complex life forms. The apparent contradiction lies in the domain in which the laws applied. Mechanical systems are closed systems isolated from the environment and their laws pertain to only a small part of the universe. In contrast, most biological and social systems are open, not closed. They exchange matter and energy with the environment.

Prigogine argued that classical thermodynamics holds in systems that are in equilibrium or near-equilibrium, such as pendulum clocks, steam engines, and solar systems. These are stable systems in which small changes within the system lead to adjustments and adaptations. They are described mathematically by the great seventeenth and eighteenth century theoretical advances in calculus and linear differential equations. But what happens when the input is so large that a system cannot adjust? In these far-from-equilibrium systems, nonlinear relationships take over. In such cases small inputs can produce new and unexpected effects.

Prigogine's far-from-equilibrium thermodynamics allows for the possibility that higher levels of organization can spontaneously emerge out of disorder when a system breaks down. His approach applies to social and ecological systems, which are open rather than closed, and helps to account for biological and social evolution. In the biological realm, when old structures break down, small inputs can (but do not necessarily) lead to positive feedbacks that may produce new enzymes or new cellular structures. In social realms, revolutionary changes can take place. On a large scale, a social or economic revolution can occur in which a society regroups around a different social or economic form, such as the change from gathering-hunting to horticulture, or from a feudal society to a preindustrial capitalist society. In the field of science, a revolutionary change could entail a paradigm shift toward new explanatory theories, such as the change from a geocentric Ptolemaic cosmos to a heliocentric Copernican universe.[7]

The recent emergence of chaos theory in mathematics suggests that deterministic, linear, predictive equations, which we learn in freshman calculus and which form the basis of mechanism, may apply to unusual rather than usual situations. Instead, chaos, in which a small effect may lead to a large effect, may be the norm. Thus a butterfly flapping its wings in Iowa can result in a hurricane in Florida. Chaos theory reveals patterns of complexity that lead to a greater understanding of global behaviors, but militate against over-reliance on the simple predictions of linear differential equations.

The butterfly metaphor originated with Edward Lorenz, Professor of Meteorology at the Massachusetts Institute of Technology, who used it to describe the phenomenon of sensitive dependence on initial conditions. In a talk entitled, "Predictability: Does the Flap of a Butter-fly's Wings in Brazil Set Off a Tornado in Texas?" he wrote: "The question which really interests us is whether . . . for example, two particular weather situations differing by as little as the immediate influence of a single butterfly will generally after sufficient time evolve into two situations differing by as much as the presence of a tornado. In more technical language, is the behavior of the atmosphere unstable with respect to perturbations of small amplitude?"

Lorenz's work, for which he won the 1983 Crafoord Prize of the Royal Swedish Academy of Sciences, led him to question the possibil-ity of finding suitable linear prediction formulas for weather forecasting and instead to develop models based on nonlinear equations. He argued that irregularity is a fundamental property of the atmosphere and that the rapid doubling of errors from the effects of physical features precludes great accuracy in real-world forecasting. Most environmen-tal and biological systems, such as changing weather, population, noise, non-periodic heart fibrillations, and ecological patterns, may in fact be governed by nonlinear chaotic relationships.[8]

In the realm of biology, Charles Birch has offered a "postmodern challenge" to the mechanistic approach of atomistic units governed by external relationships. Arguing that the dominant model of life in biology is both mechanistic and reductionist, he substitutes an "ecolog-ical model" based on internal relations. As one moves up the hierarchy from electron to atom to cell to organism, the properties at any one level do not totally predict those at the next. There are new relations between the units, not just a basic rearrangement of units. Each being is a subject that is interrelated with its environment, not something that can be studied in isolation. In the ecological model of the brain, such as that of Karl Pribram, images do not result from the addition but the interaction of many cells. If some brain cells are removed the image is reduced in clarity, but parts of it do not vanish. There is thus no point to point correspondence as in a lens, but many to one relationships, as in a hologram. In genetics, the ecological model says that genes are not like atoms or billiard balls. The way the DNA expresses itself depends on the cellular environment. Molecules and their chemical environments are in dynamic equilibrium and pathways are probable, not determined. Molecular biology is thus molecular ecology.[9]

The Gaia hypothesis of atmospheric chemist James Lovelock offers another biological challenge to the mechanistic model. Named after the Greek earth goddess Gaia, the hypothesis states that "the physical and chemical condition of the surface of the earth, of the atmosphere, and of the oceans has been and is actively made fit and comfortable by the presence of life itself." The biosphere is a self-regulating (cybernetic) system. The hypothesis challenges mechanism by offering the idea that Gaia as a living earth is more than the mere sum of its parts. Life itself plays an active role in maintaining the conditions necessary for its own continuation.

Lovelock's central idea is that "the living matter, air, oceans, and land surface form a complex system which can be seen as a single organism and which has the capacity to keep our planet a fit place for life." The atmosphere is not merely a collection of gases in more or less definite proportions, but a biological construction that is an extension of a living system, much like the hair on the back of a cat or the shell of a snail. If even small deviations from the present proportions of gases occurred, it would be a disaster for life itself. Oxygen, for example, at 21 percent of the atmosphere is the safe upper limit in which life can occur; even small increases would lead to an increase in terrestrial fires. At 25 percent the planet would be a raging conflagration extinguishing even the possibility of life.

Other atmospheric gases are maintained by life processes. Methane, produced in the muds of wetlands by anaerobic bacteria, bubbles to the surface where it combines with oxygen to produce water and carbon dioxide, thus preventing the slow build up of atmospheric oxygen concentrations. Nitrous oxide (N_2O) is produced by microorganisms in the soils and seas. It provides a counterbalance to methane and also regulates the amount of oxygen. Nitrogen, which is 79 percent of the atmosphere, is produced by denitrifying bacteria which return it to the air. Without life, nitrogen and oxygen would both return to the sea. Nitrogen dilutes oxygen, regulates combustion, and stabilizes climate. Ammonia is also of biological origin, producing rain with a pH of 8. Water, an essential, chemically-neutral substance, returns oxygen to the atmosphere and hydrogen to outer space. The entire interconnected global system of living and non-living things contains internal feedbacks that keep the chemical percentages within the ranges suitable for life's continuance. Later Lovelock, working with scientist Lynn Margulis of Boston University, extended his hypothesis to include oceans and soil.[10]

The Gaia hypothesis, however, has been criticized as being both teleological and tautological. In 1988, the American Geophysical Union held a conference in San Diego on the Gaia hypothesis that included well known scientists—skeptics who questioned the extreme purposefulness built into the hypothesis, and supporters who explored possible connections with hot springs, the human brain, and the extinction of dinosaurs. James Kirchner of the University of California sees it as a nest of hypotheses ranging from the self-evident to the highly speculative. At the straightforward end of the scale, it simply reiterates the well-documented linkages between biogeochemical and biological processes, while emphasizing the importance of feedback loops between them. At the speculative end is the more questionable concept that biological processes regulate the physical environment maintaining favorable conditions for life. The latter, Kirchner asserts, is untestable, unproveable, and unfalsifiable. Gaia is perhaps nothing but a tautology.[11]

Nevertheless, the Gaia metaphor caught on rapidly during the 1980s as a powerful new image for uniting the combined destinies of people, other organisms, and inorganic substances. Environmental historian J. Donald Hughes looked at Greek ideas of the earth as a goddess and the cosmos as an organism in his 1982 article, "Gaia: An Ancient View of our Planet." The National Audubon Society Expedition Institute sponsored a 1985 public symposium, "Is the Earth a Living Organism?" that featured papers by scientists, anthropologists, historians, poets, American Indians, and spiritualists. Feminists took up the theory as support for the ancient goddess Gaia (see Chapter 5) and opened Gaia bookstores to market goddess statues, books, and records. Musician Paul Winter composed, "Missa Gaia, A Mass in Celebration of Mother Earth," which was recorded live in the Cathedral of St John the Divine in New York and in the Grand Canyon.

The hypothesis also sparked an array of books that pictured threats to the global Gaian ecosystem, explored scientists' and economists' thoughts on its political implications, and extended the idea to the field of environmental and bioethics.[12]

These new approaches to science are consistent with deep ecology's call for a new metaphysics. They are based on a different set of assumptions about the nature of reality than mechanism—wholeness rather than atomistic units, process rather than the rearrangement of parts, internal rather than external relations, the nonlinearity and unpredictability of fundamental change, and pluralism rather than reductionism. Yet could a postclassical science embodying such a vision be socially created and accepted? If so, it might provide alternative ethical guidelines for humanity's relationship with the environment.

EASTERN PHILOSOPHY

In many ways the assumptions of the new postclassical science resonate with the much older metaphysical beliefs of ancient Asia: Taoism, Buddhism, Zen Buddhism, Hinduism, and the many sects and traditions within Chinese, Japanese, and Indian thought. Taoism offers an example of an alternative approach to knowledge, ethics, and the study of Nature.

In the sixth century B C in China, the "Old Master," Lao Tzu, set down a collection of classic aphorisms known as the *Tao Te Ching* (the Ts are pronounced as Ds), or *The Way*. A contemporary of Confucius who developed a philosophy of practical ethics, over the succeeding centuries Lao Tzu and his Taoist philosophy became associated with "the people," while Confucianism appealed more to China's bureaucratic élite. By the end of the sixth century A D, Taoism was established as a popular cult, infused with alchemy, healing, popular magic, and ultimately with scientific developments such as the magnetic compass and gunpowder. Taoist priests incorporated Buddhist teachings about the afterlife and Nirvana, or eternal happiness. Confucian scholars, who were more concerned with morals, abstract social ethics, and the practice of the good life, looked down on Taoism as a "popular" emotional religion.[13]

The Tao, or the way, is the ultimate reality, the One that underlies the appearances. As cosmic process, it is the way of the universe. Taoists emphasize changes and flows within the whole, observing patterns within the cyclic, ceaseless motion of going and returning, expansion and contraction. Human intellect can never fully grasp the Tao, but people can observe nature to discover its ways. Its nonanalytic, intuitive, scientific approach achieves insights into transformation and change, into growth and decay, life and death through observation of the natural world. Taoist method links opposites, stressing contrary aspects, innate tensions, and spontaneity. Thus yin and yang are polar opposites within constant change. Yin represents the active, yang the receptive; yin is sunny, yang is shady; yin is light, yang is dark; yin is male, yang is female; yin is firm, yang is yielding, yin is heaven, yang is earth and so on. The body is a balance between yin and yang, outside and inside, front and back. The Ch'i is its vital energy, the continuous flow that connects yang organs by way of yin meridians.

Taoist ethics say that to achieve something, one must start with its opposite. To retain, one must admit the opposite. Action is inaction. One should not force change; instead change stems from within in accordance with the flow of the Tao and the natural order. Good is balanced with bad.

Taoism is a form of dialectical idealism. Mao Zedong contrasted his own Marxian philosophy of dialectical materialism by quoting the Tao's idealist assumptions: "The Tao that can be told is not the eternal Tao; the name that can be named is not the eternal name." The particular is not the reality. The Tao is nameless. It is form without object, shape without shape. "We look at it and do not see it; its name is the invisible, the inaudible, the formless." But like Maoism, Taoism is rooted in the dialectic. The contradictions between two opposites produce change. Being and non-being produce each other. A cup is molded of clay, but its non-being, or hollow space, is the useful part. Long and short, high and low, front and back accompany each other. "To be crooked is to be perfected; to be bent is to be straightened; to be lowly is to be filled; to be senile is to be renewed; to be diminished is to be able to receive."[14]

As in the postclassical process sciences, the Tao is the world's underlying energy. "What the Tao produces and its energy nourishes, nature forms and natural forces establish. On this account there is nothing that does not honor the Tao and reverence its energy." The Tao "produces, but keeps nothing for itself; acts, but does not depend on its action; increases, but does not insist on having its own way. This indeed is the mystery of energy."[15]

What distinguishes Eastern from Western philosophies is often the use of analogy and metaphor rather than abstraction. Eastern thought resists the unifying, abstract, transcendent concepts so characteristic of western science. Rather than theory, it offers a fine-tuned image, instead of argument an inimitable experience, in place of syllogism, an evocative aphorism. Eastern ways of knowing, alternatives within the western philosophical tradition, and the postclassical sciences are some of the sources for deep ecology's challenge to the mechanistic worldview. Yet deep ecology is itself criticized for its lack of both a politics of transformation and a politics of gender difference.[16]

CRITIQUES OF DEEP ECOLOGY

"How deep is deep ecology?" asks philosopher George Bradford. Admitting that the "environmental crisis is a crisis of a civilization destructive in its essence to nature and humanity," Bradford excoriates deep ecologists for their lack of a political critique. They fail to recognize that the idea that all things in the biosphere have an equal right to exist is just as much of a projection of human sociopolitical categories onto nature as is the anthropocentrism they criticize. They fail to extend the ecological concept of interrelatedness to technology as a system or to the extractive empire of global capitalism. They take the character of capitalist democracy for granted rather than submitting it to a critique.

Deep ecologists who make a sharp distinction between wilderness and anthropocentrism fail to consider that humans are also animals.

Moreover they fail to recognize the ethnocentrism of their own concept of wilderness as devoid of human presence, especially that of aboriginal peoples who for thousands of years inhabited the very lands they now wish to define as wilderness.

Many deep ecologists accept the Malthusian premise that the root of the problem is too many people. Catton, for example, premises the need for a New Ecological Paradigm on the fact that human numbers have exceeded the carrying capacity of the environment. "Population growth," retorts Bradford, "is certainly a cause for concern. . . . More than 900 million people are presently malnourished or starving, and hunger spreads with rising numbers. But Malthusian empiricism sees many hungry mouths and concludes that there are too many people and not enough resources to keep them alive." Instead of seeing scarcity and famine as "inevitable, irrevocable, even benign,"[17] hunger is the product of maldevelopment and requires social rather than deep ecological transformation.

A second critique focuses on the socioeconomic and scientific *naiveté* of deep ecology. Capra's approach in particular, says Stephan Elkins, idealizes culture as the reflection of a society's values and the key to action. Far from examining the ways in which values are related to social structures or analyzing how social structures might change, Capra simply assumes that values and worldviews change over time following cyclical patterns of genesis, growth, maturation and decline. Minority groups with new ideas appear (such as feminists, Greens, and bioregionalists), the old socioeconomic forms disintegrate, and a new cycle begins, as in the current transition to a non-patriarchal, solar age. Change is painless, benign, and independent of political struggle.

Instead, argues Elkins, values emerge from people's everyday experience as formed by their place in class society, not from learning about the Cartesian-Newtonian worldview or the new ecological paradigm. Problems of economic production stem not from our culture's obsession with growth, the desire for indefinite expansion, and consumer inducements to buy and throw away, but from the unequal distribution of private property.

Capra advocates the systems approach of the new ecological paradigm, presenting it as an objective reflection of the systemic wisdom of nature that can provide balance and harmony for society. But science for Elkins is a social product: "A society's view of nature must be seen as part of its self-interpretation, reflecting social relations and its relation to the natural environment." No less than the mechanistic worldview, which Capra criticizes, systems theory is equally reductive, selecting ecological relations as the functions that science mathematically describes. Science, Elkins argues, is thus magically transformed from an inhibiting mechanistic rearguard into a revolutionary life-affirming vanguard. From a force of destruction and domination, it suddenly becomes a source of hope and salvation. The systems-theoretical core of Capra's ecological paradigm could be appropriated, not as a source of cultural transformation, but as an instrument for technocratic management of society and nature, leaving the prevailing social and economic order unchanged.[18]

A third critique of deep ecology comes from ecofeminism (see Chapter 8). In "Deeper than Deep Ecology: the Eco-Feminist Connection," Ariel Kay Salleh offers a critique of a critique. Philosopher Arne Naess' use of the generic term "man" in his 1972 paper is more than a semantic or sexist flaw. Although Naess promotes biospheric egalitarianism and a "relational total-field image" (see Table 4.1), he and other deep ecologists fail to see the historical and philosophical connections between the domination of nature by "man" and the domination of women by men. "The master-slave role which marks man's relation with nature," argues Salleh, "is replicated in man's relations with woman." The "anti-class" posture offered by Naess is superficial, ignoring the connection between nature as commodity and woman as commodity in patriarchal society. Moreover, the artificial limitation of the human population advocated by deep ecologists in order to achieve species equality is rationalist and technist. This approach, according to Salleh, contradicts the life-affirming values of both deep ecology and woman as bearer of life.[19] Finally, many ecofeminists argue that deep ecology's anthropocentric critique ignores androcentrism—it is men not women who in fact have historically created and controlled science and technology—and gender difference— women lose identity in merging with the larger ecological self.

Could deep ecology be cured of its antifeminist bias through greater sensitivity to its own language and analysis? The answer is no. This would be a mere bandaid. An even deeper social feminist critique exposes the biases in both patriarchy and capitalism. The hegemony of capitalists over laborers depends on the exploitation of nature as a free gift to capital. The hegemony of men over women is necessary to maintain women's double "second-shift" labor in the home and the workplace, whether in capitalist or state socialist societies. A science rooted in the twin assumptions of atomism and objectivity legitimates the domination of both nature and women.

Mechanistic science is patriarchal inasmuch as it has been historically dominated by men who have produced "truths" about reality. The result is dualistic thinking in which the world is interpreted in terms of dominance and submission, objectivity and subjectivity, rationality and emotion, with the first characteristic of each pair being associated with men and the second with women. Women have not participated in the scientific and cultural projects that have defined women's "nature" as emotional, unruly, and subjective and men's "minds" as rational, unbiased, and objective—the epitome of science itself.[20]

RECONSTRUCTIVE SCIENCE

Could there be a science that would be consistent with egalitarian and feminist social values? Much of nineteenth and twentieth century science was influenced by the logical positivist philosophy that mathematics and experimentation lead to certain knowledge of an external real world. Historians and philosophers of science in the late-twentieth century, however, have questioned this positivist approach. Thomas Kuhn's *Structure of Scientific Revolutions* (1962), raised two fundamental questions: Does each age construct its own scientific reality? Does science actually represent progress in the objective knowledge of nature? During the past decade a social constructivist philosophy of science has responded that science is basically a social construction by groups of scientific practitioners who have access to the corporate and governmental systems of power that review proposals and fund research. This school argues that what counts as scientific knowledge is based on: (1) the acceptance by a community of practitioners of what counts as a scientific "fact," (2) the social selection and deselection of facts and theories that interpret natural phenonmena, and (3) their inscription into texts accepted as state of the art knowledge by the scientific community and taught to the next generation of scientists. These considerations raise even more radical questions:

1. Can there be a pristine scientific knowledge beyond social, gender-biased, and value-laden processes of scientific investigation and systems of institutional support?
2. Can there be a reconstructed postclassical science and a reconstructive way of knowing nature?
3. Can there be a reconstructed system of knowledge consistent with egalitarian, democratic values that would lead to a sustainable ecology and society in the twenty-first century?

In their book *New Ways of Knowing*, Marcus Raskin and Herbert Bernstein offer a manifesto of reconstructive knowledge. "The world—that is, the world we communicate about—is transformed by description of it. Knowledge workers shape the social organization in which our inquiries about nature take place. And our cognitive understandings of the world are manufactured, indeed, usually *man-ufactured*." A reconstructive knowledge method will be dedicated to the social good, concern with public participation, and the incorpora-tion of humane values into research goals. It starts with choosing a research topic, a small interdisciplinary research group to work on it, and a day-to-day method that is guided by future moral applications. Questions and answers should be based on social realities, not on disciplinary inquiries. Small groups of researchers from several fields should thoroughly discuss the social and ecological implications of their own projects before undertaking them. Research that denies humanity a future (such as chemical and biological weapons) should not be funded or pursued. Instead research programmes that lead to an improvement in the quality of life for disadvantaged groups and the restoration of diversity to the natural world will have priority.[21]

Feminists such as Ruth Bleier, Evelyn Fox Keller, and Sandra

Harding likewise emphasize a reconstructive knowledge based on principles of interaction (not dominance), change and process (rather than unchanging universal principles), complexity (rather than simple assumptions), contextuality (rather than context-free laws and theories), and the interconnectedness of humanity with the rest of nature. An ecological approach to problem-solving would be based on human interactions with the nonhuman world, recognition of the imbeddedness of humans in complex biological and social proceses, and the context dependence of particular ecosystems in particular times and places. Such a vision of science could contribute to a new relationship with the natural world because it would place humans within it rather than dominant over it and recognize women's roles in the reconstruction of knowledge.[22]

CONCLUSION

The ideas of deep ecology, alternative philosophies, the emerging postclassical sciences, feminism, and reconstructive knowledge point to the possibility of a new worldview that could guide twenty-first century citizens in an ecologically sustainable way of life. The mechanistic framework that legitimated the industrial revolution with its side-effects of resource depletion and pollution may be losing its efficacy as a framework. But a nonmechanistic science and an ecological ethic must be consistent with a new social ecology (see Chapter 6) and with feminist values (see Chapter 8). It must support a new economic order grounded in the recycling of renewable resources, the conservation of nonrenewable resources, and the restoration of sustainable ecosystems that fulfill basic human physical and spiritual needs (see Chapter 9).

Deepest ecology is both feminist and egalitarian. It offers a vision of a society that is truly free. It recognizes that nature is a social construction that changes over time. People have the power to construct nature as a free, autonomous subject, not a dominated object— a nature that is an equal partner with equal women and men. Deepest ecology also recognizes that science is enmeshed in socially negotiated relationships with nature, relationships that respond to the needs of society. Which research projects are selected and funded depends on social goals; which relations are codified by science depends on social needs. If social goals start with the fulfillment of basic human and quality-of-life needs, then people working together through social movements can create a truly egalitarian, ecological society. Perhaps then Nature as equal partner can be healed.

FURTHER READING

Allaby, Michael. *A Guide to Gaia: A Survey of the New Science of Our Living Earth*. New York: Dutton, 1989.

Bleir, Ruth. *Science and Gender: A Critique of Biology and its Theories on Women*. New York: Pergamon, 1984.

Bohm, David. *Wholeness and the Implicate Order*. Boston: Routledge and Kegan Paul, 1980.

Bradford, George. *How Deep is Deep Ecology?* Hadley, Ma.: Times Change Press, 1989.

Briggs, John P., and F. David Peat. *Looking Glass Universe: The Emergence of Wholeness*. New York: Simon and Schuster, 1984.

———. *Turbulent Mirror: An Illustrated Guide to Chaos Thoery and the Science of Wholeness*. New York: Harper and Row, 1989.

Callicott, J. Baird, and Roger T. Ames. *Nature in Asian Traditions of Thought* . Albany, N.Y.: State University of New York Press, 1989.

Capra, Fritjof. *The Tao of Physics*. Berkeley, Ca.: Shambala, 1975.

———. *The Turning Point*. New York: Simon and Schuster, 1982.

Devall, Bill and George Sessions. *Deep Ecology: Living as if Nature Mattered*. Salt Lake City: Peregrine Smith Books, 1985.

Fox, Warwick. *Toward a Transpersonal Ecology: Developing New Foundations for Environmentalism*. Boston: Shambala, 1990.

Gleick, James. *Chaos: The Making of a New Science*. New York: Viking, 1987.

Griffin, David, ed. *The Reenchantment of Science: Postmodern Proposals*. Albany, N.Y.: State University of New York Press, 1988.

Harding, Sandra. *The Science Question in Feminism*. Ithaca: Cornell University Press, 1986.

Keller, Evelyn Fox. *A Feeling for the Organism: The Life and Work of Barbara McClintock* (New York: W. H. Freeman, 1983).

————. *Reflections on Gender and Science* (New Haven: Yale University Press, 1985).

Kuhn, Thomas. *The Structure of Scientific Revolutions*. Chicago: University of Chicago Press, 1970.

Lao Tzu. *The Tao-Teh King*, trans. C. Spurgeon Medhurst. Wheaton, Ill.: Theosophical Publishing House, 1972.

Lawrence, Joseph. *Gaia: The Growth of an Idea*. New York: St. Martin's Press, 1990.

Lovelock, James. *Gaia: A New Look at Life on Earth*. New York: Oxford University Press, 1979.

————. *The Ages of Gaia: A Biography of Our Living Earth*. New York: W. W. Norton, 1988.

Matthews, Freya. *The Ecological Self.* London: Routledge, 1991.

Miller, Alan. *Gaia Connections: An Introduction to Ecology, Ecoethics, and Economics*. Savage, Md.: Rowman & Littlefield, 1990.

Myers, Norman., ed. *The Gaia Atlas of Planet Management*. New York: Doubleday Anchor, 1984.

Naess, Arne. *Ecology, Community, and Lifestyle*. Cambridge: Cambridge University Press, 1989.

Needham, Joseph. *Science and Civilization in China*, vol. 2. Cambridge: Cambridge University Press, 1956.

Prigogine, Ilya. *Order Out of Chaos: Man's New Dialogue with Nature* New York: Bantam, 1984.

Raskin, Marcus G. and Herbert J Bernstein. *New Ways of Knowing: The Sciences, Society, and Reconstructive Knowledge*. Totowa, N. J.: Rowman and Littlefield, 1987.

Rosser, Sue. *Female Friendly Science*. New York: Pergamon, 1990.

Thompson, William Irwin, ed.. *Gaia: A New Way of Knowing*. Great Barrington, Ma.: Lindisfarne Press, 1987.

Tobias, Michael, ed., *Deep Ecology*. San Diego: Avant Books, 1985.

Tuana, Nancy, ed. *Feminism and Science*. Bloomington, In.: Indiana University Press, 1989.

5

SPIRITUAL ECOLOGY

People are sitting in a circle in a woodland clearing, warm earth below, blue sky above, sun shining through leaves and pine needles. They have just returned from special individual places and have taken on the identities of other natural beings. With paper and paste, colored pens and scissors, they make masks. Passing a smoking shell and a bowl of fresh water, they begin the ritual.

Turning first to the east, then to the other three directions, they invoke the powers of nature. They invite the beings of the Three Times—naming those who have nurtured the earth before, those who are saving it in the present, and those of future times for whom the earth is being preserved. Each being in turn speaks for itself and its kind, telling of its place in the earth's order. "I am rainforest; I am kangaroo; I am mountain; I am lichen." Then a few remove their masks and move into the circle's center to listen as humans to what is happening to the others.

> I am rainforest. . . . You destroy me so carelessly, tearing down so many of my trees for a few planks. . . . You cause my thick layer of precious topsoil to wash away, destroying the coral reefs that fringe me. . . . Your screaming machines tear through my trunks, rip my flesh, reducing me to sawdust and furniture.

After all have listened, a human finally speaks. "We hear you fellow beings. We feel overwhelmed. We need your help. Are there powers and strengths you can share with us in this hard time?" Each being offers help and shares its gifts with the others, leaving its mask and joining the humans in the center. The humans join together, humming as one organism, then break apart with singing and dancing.[1]

THE COUNCIL OF ALL BEINGS

A ritual of despair and empowerment, the above Council of All Beings was developed by Joanna Macy, John Seed, and others to help people find and act on their own powers to save the planet. From its origins in Australia's movement to save its rainforests, the ritual Council has spread around the world to Tibet, England, California, and onward. The rituals are not intended as a subsitite for social action, but as preparation for it. They bring to consciousness the natural history of the planet and convey an authority to act on its behalf. Identification with the earth and its beings empowers each person and removes doubts and hesitations.

Spiritual ecology, like deep ecology, is a product of a profound sense of crisis in the ways that twentieth century humans relate to the environment. Like deep ecology it focuses on the transformation of consciousness, especially religious and spiritual consciousness. Recognizing the importance of some form of religious experience or worship in the lives of most people, spiritual ecologists attempt to develop new ways of relating to the planet that entail not an ethic of domination, but one of partnership with nature. Religious ideas create strong moods and motivations that act as an ecocentric ethic, guiding individuals and social movements towards new modes of behavior. The ideas of spiritual ecologists thus motivate individuals active in green ecological and ecofeminist social movements (see Chapters 7 and 8). Through rituals, a sense of reverence for nature can arise, centering people for social action.

Other rituals reinforce Macy and Seed's Council of All Beings approach. Gaia meditations call upon people to participate in the cycling of the ancient elements—earth, air, fire, and water—through their bodies and lives. Just as water pours in and out of the body and its fluids, so it flows through the earth's springs, rivers, clouds, and rain. Earth, rock, and soil find their way into the body's molecules and cells and they in turn become ashes and dust. As air is inhaled and exhaled it takes from and gives back to the trees and plants the sustenance necessary for life to continue. The sun's fire, the body's heat, and the cosmic big bang are the same changing manifestations of matter and energy. Each person is part of the long unbroken chain of creation. Consciousness of that history and interconnectedness reinforces belonging and gives strength to act.[2]

In one and two day workshops people engage in these ritutals and share ecostories of times when they felt the power of the natural world or pain on what is happening to it. They honor endangered species in a "bestiary" mourning, calling out the names of the species leaving the planetary family forever. During ecomilling, they dance and move in silence, looking, touching, and encountering each other in all their personal vulnerability to the poisoning of the planet and their personal power to heal it. At the end of the workshop people share their reflections and plan subsequent actions and meetings.

Joanna Macy's empowerment workshops are based on a fivefold spiritual response to the pain that so many people feel about the two major threats to the planet: the possibilites for nuclear holocaust and for ecological crisis. The principles on which her work is based are:

1. Feelings of pain for our world are natural and healthy.
2. This pain is morbid only if denied.
3. Information alone is not enough.
4. Unblocking repressed feelings releases energy and clears the mind.
5. Unblocking our pain for the world reconnects us with the larger web of life.

It is through awareness of our human capacity to suffer with the world that we experience dimensions beyond ourselves, and through this ongoing awareness grasp the power to heal. "Moving through our pain for our world," she states, "is no more our doing as separate egos, than childbirth is the doing of the mother. For it is the deep ecology of life itself, if we let it, that draws us home to the awareness of our true nature and power." Because she believes that our generation's crime against the future is so terrible, Macy proposes that earth burial sites filled with toxic and irradiated materials need to be consecrated as guardian sites where the containers are religiously monitored and repaired. Much like the communities who have camped at the sites of United States nuclear bases, dedicated surveillance communities must continuously remind us and our descendants of the crippling power of these materials for millennia to come.[3]

NATURE SPIRITUALITY

Do women need the goddess? A resounding, "yes," say many feminists and devotees of new age spirituality. The goddess is an important replacement for the patriarchal symbolism of a male God, the power of which permeates all our cultural institutions, even non-religious ones. "Religions centered on the worship of a male god," says Carol Christ, "keep women in a state of psychological dependence on men and male authority, while at the same time legitimating the political and social authority of fathers and sons in the institutions of society." For women, the goddess is an affirmation of female power and female relationships. She symbolizes their importance as bringers forth of life and their connections with the earth.[4]

A new iconography emerging from festivals, workshops, and conferences brings women in touch with submerged feelings that unite them with the powers of nature. The goddess has become a source of inspiration to female artists, musicians, poets, and actors seeking ways to reimage and heal human relationships with nature. Goddess rituals celebrated at the solstices and equinoxes enhance the personal meanings attached to cycles of life and death, menstruation and menopause.

Men too are acknowledging the need for nature symbols in their lives. Many men find in rituals an affirmation of their own connections to nature and an ethic of caring for the earth. In consciousness raising workshops, men renew their spiritual relationships with nature through taking on the identities of figures of the forest such as the horned god, symbol of a generative creative force in nature—the Greek Pan, the Green Man of Europe, Pan Robin of the Green, the magician Merlin—and through taking on the identities of animals. San Francisco's Harvey Stein invites men to "climb in the body of Geb," Egyptian god of the earth, "live the archetypes of Dionysus the Ecstatic, the Wild Man of the Forest, the Lord of the Animals," and through the Green Man of Europe with his leafy face, to "feel tree and animal life in our bodies." He suggests that patriarchy is oppressive to men as well as women and that men can offer strength and tenderness both to each other and the earth.[5]

Men's movement gurus such as Robert Bly, James Hillman, Robert Moore and Shepherd Bliss, inspired by mythological meanings in the work of Mircea Eliade, Joseph Campbell, and Carl Jung, facilitate male encounters with their "deep male" selves. Bly believes that men need to get in touch both with their feminine side and the deeper "wild man" within. Moore sees a need to promote a planetary vision, confront gender antagonisms, and reconcile masculine and feminine in mutual empowerment and cooperation. For Bliss, Orpheus is a male symbol of an earth-dwelling sprituality who contrasts with the transcendent sky gods of Olympus. Using lyre and song, rather than the blade and sword, Orpheus symbolizes gentleness and persuasion, love of beauty, and deep connections between men and nature. In rituals held in caves and woodlands, men (women may also be included) "descend in search of the deep feminine, singing to the goddesses, and we ascend to return to an earthy masculinity to guide us during these turbulent times."[6]

The current earth-based spirituality movement is part of an explosion of research on ancient nature religions. Archeologist Marija Gimbutas contrasts the goddesses and gods of southeastern old Europe during the period 7000 BC to 3500 BC with the sky gods brought by waves of horse-mounted Kurgan invaders from the Eurasian steppes between 4400 B. C. to 2800 BC. The horticulturalists of old Europe were settled, seemingly peaceful, bands whose life cycle focused on birth, death, and regeneration rituals centered on the female principle. They produced statues of seated goddesses with large bellies, buttocks, and cylindrical necks, woman-bird hybrids, and bird masks. Hybrid male-female and human-animal figures indicate a fusion with rather than a dualism between humans and nature. Other cosmological images, found on vases, lamps, altars, and walls, include spiral snake designs (symbolizing regeneration though skin shedding), "cosmic" eggs with snakes wound around them, fish designs, water birds, butterflies, and bees.[7]

Throughout the ancient world, female deities were worshipped as bringers of natural fertility and were often found in association with male gods. In Mesopotamia, the female fertility goddess Ishtar (Inanna) was worshipped during prehistory, but with the introduction of agriculture and domesticated animals, she was accompanied by her son-lover Tammuz. As the generative power in nature, Ishtar renewed life each spring, descending into the underworld to bring back her dead son Tammuz. In Egypt, Isis was the symbol of the maternal principle who produced vegetation through impregnation by the sun god, her brother and spouse, Osiris. Every spring her tears overflowed, producing the flooding of the Nile. Her flowing gown was decorated with stars and flowers. In one hand she carried a pail, symbolic of the flooding of the Nile, while the other shook the sistrum, a rattle which continually agitated the powers of nature. In Greece, fertility rituals were centered on Demeter (the Roman Ceres) and her daughter Persephone (Roman Proserpina). Celebrated at Eleusis, the rites reenacted the abduction of Persephone by Pluto and the wandering of the grief-stricken Ceres during the four months each year that her daughter was lost to the underworld.

Art historian Pamela Berger has traced, through art imagery, the transformation of the goddess from Graeco-Roman protectress of the grain to medieval saint in her book *The Goddess Obscured*. Demeter is depicted with serpents around her arms, holding stalks of wheat. "In ancient Greece Gaia [was] syncretized with Demeter goddess of grain who created plant life, conserved it, and dissolved vegetation in order to renew it." The Roman Terra Mater, shown on a first century breast plate as a mother goddess with cornucopia, grain stalks, and children in her lap, appears on a ninth century bookcover with flowing hair, supporting a cornucopia, and welcoming children. In the eleventh century, she appears nursing a serpent and cow and again with Adam and Eve on her lap with the snake as the serpent of Eden. In the grain miracle stories, she has been transformed into a saint, protecting the harvest from evil and miraculously causing grain to ripen as she passes. Finally, Mary replaces Demeter as grain protectress.[8]

Images such as these have inspired women artists and performers in the late twentieth century. In *The Once and Future Goddess*, Elinor Gadon skillfully juxtaposes a color plate of the medieval Tellus Mater with Meinrad Craighead's 1980 colored ink drawing of Mother Earth with flowing hair, animals and humans nestled at her feet, offering fruits from her garden. She shows the Stone-Age large-breasted Earth Mother of Willendorf next to a colored photograph of a 1985 perfor-mance by Susan Maberry as the earth mother on the day after the nuclear holocaust. The multi-breasted Artemis from the first century is placed beside an illustration of Louise Bourgeois as Artemis from the 1980 performance of "A Banquet/Fashion Show of Body Parts."[9]

Jewish women have found spiritual empowerment in a revival of God the Mother as an aspect of the divine. The Shekinah is the female spirit of God whose presence dwells in human beings. The importance of the shekinah was recognized in the writings of Jewish rabbis during the exile by the Romans in the first century of the Christian era. It then went underground until the twelfth century when it was revived in the Kabbalah, a mystical form of the Jewish religion. Jewish artist Gila

Yellin Hirsch of Los Angeles depicts her power in paintings entitled Shekinah (1976) and Emergence (1981), while Beth Ames Swartz, who travelled to Israel to visit sacred sites of Jewish females, painted The Red Sea (1983) in honor of Moses' sister Miriam.[10]

The presumed dominance and subsequent decline of ancient goddess symbols and nature spirituality in western culture have political implications. Some feminists have used archeological and mythological evidence to argue that societies in prehistory may have been matriarchal, that is, under female political rule. In *The Chalice and the Blade*, however, Riane Eisler uses the same evidence to make a case for dominator versus partnership societies. In her view, matriarchy and patriarchy are both examples of the dominator model, symbolized by the blade, in which the ranking of one sex is higher than that of the other. The partnership model, symbolized by the chalice, is based on linking, rather than ranking, and offers hope for an egalitarian political and economic society in the future.

Using both feminist theory and cultural transformation theory, Eisler argues that an original partnership society in prehistory took a 5000 year detour into a dominator society. Yet a future society based on a partnership model between women and men and humans and nature may be emerging. In this society the "androcratic virtues" associated with the domination of nature and other peoples will be replaced by "gylanic consciousness." Gylany is derived from the Greek roots, "gyne" meaning woman and "andros" meaning man, linked by the letter "l" from the Greek word "lyein" meaning to resolve or to set free.[11]

THE OLD RELIGION

They gather on hilltops and beaches, in groves and fields, in rented storefronts and condominium penthouses to celebrate the full moon. Taking hands they cast the circle around an altar of flowers and candles, breathing, humming, and moving together to raise power, share it, and then earth it. In turn they face each of the four directions, calling on the goddesses of every tradition to be with them. To the beat of a drum, moving as one long snaking form, they reenact the sacred spiral dance. Some are naked, some remain clothed. The women leave centered and renewed with the energy needed to carry on the ecological work of healing the earth.

Pagan spirituality, or the Old Religion, has been revived in modern times. Wicca is not harmful, black magic, but healing, centering power. To witches such as Starhawk and Margot Adler, magic means calling forth the power within, or the art of changing consciousness. A witch bends or shapes the unseen into new forms. The spiritual is the power and the will to change one's own life. To Z. Budapest, women are witches by right of being women. No further initiation is needed. Women form covens for support and consciousness raising— the Honeysuckle Coven of Starhawk, the Susan B. Anthony Coven of Z. Budapest, or the Compost Coven, a men's group. Covens are usually all female, but some are mixed and a few are for men only. Leadership in the covens comes from within each person rather than from power over others. Each develops her or his own inner strength. Many see these rituals as empowerment for political and social change.[12]

Whether as wicca, healing witchcraft, the religion of the Celtic druids, or as magic, as many as 100,000 people in the United States may practice a form of nature religion, animism, or pantheism based on an alive presence within nature. Ecologically oriented groups often use the lunar or pagan calendars for their gatherings and newsletters. The Elmwood Institute in Berkeley California, dedicated to promoting deep ecology, holds new moon gatherings and publishes its newsletter at the equinoxes and solstices. Earth First!, an activist group issues its newspaper eight times a year in accordance with the pagan nature holidays: Samhain (November 1), Yule (December 21), Bridgid (February 2), Eostar (March 21), Beltane (May 1), Litha (June 21), Lughnasadh (August 1), and Mabon (September 21).

Practitioners of the Old Religion have used rituals and magic in political demonstrations. For example, members of Starhawk's Matrix affinity group protested continuation of research on nuclear weapons at California's Lawrence Livermore Laboratory. Part of a large nonviolent protest in which thousands of people were mobilized by the

Livermore Action Group during the 1980s, the members participated in training sessions in methods of nonviolent resistance. At the June demonstrations, held on the day of the summer solstice, each affinity group of six to eight people joined arms to block an entrance to the laboratory while other protesters urged employees to show support by not going to work. In the 1982 demonstration, members of Matrix created a large web, symbolic of the web of life as well as the power of women and witches. Using chants, spells, and rituals they wove yarn into a large web and imbedded it with flowers, seeds, and photographs. When a bus bearing workers approached, they used the web to blockade the road. As some members of the group were being arrested, others tied the web to the fence. In the 1983 action, one thousand arrestees were held for fifteen days in a large tent while they negotiated the terms of their arraignment and sentencing with authorities. The affinity groups, all trained in nonviolence, operated by a process of consensus decision-making that was energized and unified through rituals led by Matrix.

Yet the use of goddess spirituality and wicca in radical politics has been criticized. The rituals and meditations, crystals and pentagrams, chanting and drum-beating used at ecological conferences and demonstrations to energize and raise group consciousness are ineffective in dealing with the serious ecological problems facing the planet. Religion is a matter of individual choice and can inspire both personal transformation and political action. But when "spirituality" itself becomes a political principle, objects social ecologist Janet Biehl, and is held out as "a key to a better life," it must be scrutinized like any political platform. "A critical analysis of goddess-worshipping spirituality . . . must address not only the content of the specific myth being generated, but also the function of myth as such in an advanced industrial capitalist society."

Moreover, the archeological evidence used by Gimbutas, Eisler, and others to reconstruct goddess-worshipping egalitarian societies in prehistory, argues Biehl, "follows a simplistic philosophical idealism— namely, that cultural symbols determine social realities, confusing religious symbols with religious institutions. They fail to grapple with the question of whether an all pervasive religious authority is really a desirable alternative to a secular society." Some archeologists question the argument that the neolithic culture of Old Europe was changed by a single cause—migration of another people into the area. Others ask whether the expansion of the agrarian neolithic culture was necessarily peaceful, given the existence of arrowheads that could have been used against people as well as animals. Still others criticize the generalizations on which the arguments for mother goddess worship in prehistory are based. Of the identifiable statues and images in prehistoric art, some 35 percent are female, about 15 percent are male, and the rest are unidentifiable or simply anthropomorphic. While some female images are buxom or pregnant; others are extremely slender. Such observations undercut the presumed universality of the female fertility image.[13]

NATIVE AMERICAN LAND WISDOM

When I was small, my mother often told me that animals, insects, and plants are to be treated with the kind of respect one customarily accords to high-status adults. 'Life is a circle, and everything has a place in it,' she would say. That's how I met the sacred hoop.

So writes Paula Gunn Allen, a Keres Pueblo Indian, of the ways of native American women in her book on *The Sacred Hoop*. Many native American tribes were gynocratic, matrifocal, and matrilineal and believed that they were descended from female creator spirits— Grandmother Spider, Spirit Woman, Grandmother Woodchuck, Thought Woman, and so on. Stories passed from mother to child over the generations, taught native Americans to respect the earth and the entire animate world.[14]

In *The Ways of My Grandmothers*, Beverly Hungry Wolf speaks of the spiritual significance of the annual June Sun Dance camp of the Blackfoot tribe. Awakened each morning in tipis by an old person singing, each grandmother greets the rising sun, calling out the names of her children, grandchildren, great-grandchildren, and friends. She gives thanks for the good things of the past winter and asks for peace for the future. Learning the old ways of the grandmothers means learning which wood burns best, what meat is best to roast, how to dry it properly, how to sew lasting moccasins, and how to put up and heat tipis. It means appreciating the time when women walked long distances carrying loads of firewood and men spent countless freezing days and nights hunting for food to bring home. The Sun Dance camps grow larger every year as younger people discover spiritual strength in the older traditions.[15]

A generation of ecologically conscious people have found inspiration in native American beliefs that nature is alive and the earth is a mother. In contrast to western dualistic philosophies, most native peoples saw no distinction between animate and inanimate, natural and supernatural, body and spirit. The entire natural world was enspirited and sensate. Different entities had differing amounts of power and therefore needed to be treated with respect. From this basic assumption followed certain moral rules for treating nature. Animals, plants, and rocks needed to be addressed respectfully, and use of their names had restrictions. When killed for food, proper spiritual preparations and propitiations had to be made, the capture had to be painless, and the skinning and disposal of the remains done with respect through ritual processes.[16]

Indian orators such as Smohalla of the Columbia Basin tribes, Chief Luther Standing Bear and Black Elk of the Ogalala Sioux, and Chief Seattle of the Suquamish tribe in the Puget Sound area of the present state of Washington have preserved an earth ethic from the past that many people wish to reclaim for the future.[17] The words of Chief Seattle seem to contain the essence of the distinction between the modern American and native American land ethics:

Every part of this earth is sacred to my people. Every shining pine needle, every sandy shore, every mist in the dark wood, every clearing and humming insect is holy in the memory and experience of my people.

We know that the white man does not understand our ways. One portion of land is the same to him as the next, for he is a stranger who comes in the night and takes whatever he needs. . . . He treats his mother, the earth, and his brother, the sky as things to be bought, plundered, sold like sheep or bright beads.

Chief Seattle's words, immortalized in the movie, *Home*, and repeated in hundreds of books, articles, classrooms, and student papers represent an inspiration to return to a sane, respectful way of living within nature rather than against it.

Yet these words seem not to have been spoken by the great chief after all, but are a third- or fourth-hand version of an oral address delivered by Seattle in 1854, translated on the spot, by an unknown person, from Suquamish into English to Henry A. Smith, M.D. who in 1887 reconstructed it from extensive notes. Smith's version was later rendered into "better" classical English by William Arrowsmith and then rewritten by Ted Perry as a film script for *Home* produced in 1972 by the Southern Baptist Convention. Many of the words which resonate with modern ecological consciousness are not the original words, but contain phrases and flourishes designed to appeal to ecological idealism and the Christian religion.[18]

Does the shock of such a discovery mean abandonment of native American land wisdom? Does the argument that native peoples used cliff drives and fire drives in prehistory and guns, snowmobiles, and outboard motors in modern times mean that native Americans never had or readily abandoned an ethic of respect for nature? Were they propitiating nature out of fear rather than care for the land? No, argues philosopher J. Baird Callicott.

> If some traditional American Indian peoples practiced conservation complemented by a land ethic and maintained a long term balance between themselves and nature, then in [the words of Richard Nelson], 'If they can do it, so can we.' Their example represents hope. It also represents a role model.[19]

MAINSTREAM RELIGIONS

Mainstream churches have engaged in a variety of activities that both reinterpret the ecological crisis in spiritual terms and attempt to change society through conferences, publications, and projects. Among the Christian denominations with environmental projects are the World

Council of Churches, the American Baptist Churches, the United Methodist Church, the Lutheran Church, the Christian Church (Disciples of Christ), the Mennonite Central Committee, the Presbyterian Church, the Reformed Church in America, the United Church of Christ, the Religious Society of Friends (the Quakers) and others.

In addition, a number of seminaries, divinity schools, and universities sponsor projects and publish newsletters. These include the Commission on Stewardship of the National Council of Churches of Christ, the Eco-Justice Project of the Center for Religion, Ethics, and Social Policy at Cornell University (publishers of *The Egg* newsletter), the "Cry of the Environment" project of the Center for Ethics and Social Policy in Berkeley, California, the Friends Committee on Unity with Nature in Santa Rosa, California, the North American Conference on Christianity and Ecology in Washington D. C. and San Francisco, and the Fellowship in Prayer's Worldwide Day of Prayer and Meditation to Help Heal Mother Earth (Princeton, New Jersey).

Christian ecology sees a responsibility to reinterpret the mandate of Genesis I: 28 to "be fruitful and multiply and replenish the earth and subdue it" as the responsibility to give back to the land whatever is taken from it. This means that the nonrenewable metals should be recycled, that trees should be replanted, and that soil should be conserved. Dominion over the land means that a responsible Christian will care for the land with vision, mercy, benevolence, and compassion. Genesis 2 assigns humanity the responsibility to "dress and keep" the garden. In bringing the fruits of the garden to completion, people must renew the garden and resist the forces that despoil it. The covenant made with Noah was a covenant made with all living things. "Covenant, then includes an all-encompassing respect for the animal and vegetative life of the world, not only because they are created by God, but because they embody something of the divine nature." Stewardship means that humans have a responsibility to take care of the earth and to insure that all its beings function together in an integrated way.[20]

The Eco-Justice Project of the United Methodist Church sets out specific principles for Christian stewardship of the environment. It urges its members to promote government and community efforts to use natural resources responsibly through recycling and conservation and to allocate sufficient funds for reducing the production of toxic and hazardous chemicals, air pollutants, pesticides, and herbicides. It encourages careful stewardship of topsoil, conservation of wetlands, forests, and wildlands, maintenance of the diversity of life, and the preservation of groundwater sources. It urges the ethical and environmental examination of all new technologies and opposes the development of military weapons that would threaten the planetary environment.[21]

Other mainstream religions have also searched their great books for spiritual guidance in dealing with the ecological crisis. Ecological Judaism unites the principle of peace, Shalom, with righteousness, Tzedek. The Tikkun Olam, or the edict to heal the world is extended to repair and heal the environment. The Kosher laws of eating could be extended to forbid eating food whose production is harmful to people, animals, or the land. The holy days and Torah can be used to remind people of their interdependence with nature. "At Pesach we count the Omer, reminiscent of the ripening barley. At Shavuot we celebrate the grain harvest; at Succot the vegetable harvest." The Tu B'Shvat, or autumnal holiday of the trees can be celebrated as a major environmental holiday.[22]

For Muslim believers:

Islam . . . affords a luminous example of the centrality of ecological consciouness embedded in its inalienable view of man as the Viceregent of God on Earth. The Qur 'an teaches that the cosmos, nature, and the environment is full of signs of the Creator. . . . No religion on earth is so clearly vocative against destruction of domestic and wild life and against decimation of the God-granted natural wealth.[23]

ECOLOGICAL CREATION SPIRITUALITY

"Mother Earth in all her agony," proclaims Dominican priest Matthew Fox, "is literally crying out to the heavens themselves, as we see in the disappearance of the ozone layer over the Antarctica." Fox, founder of the Institute for Culture and Creation Spirituality in Oakland California has reclaimed a form of western mysticism and dedicated it to working for social justice and the environment. He sees his Creation Spirituality as a liberating form of worship for the First World akin to Latin America's Liberation Theology. It unifies body and spirit, joins science, art, and cosmology, frees peoples from sexism and racism, and liberates the earth from anthropocentrism.

Fox's ecological spirituality is rooted in the mystic writers of the eleventh through thirteenth centuries in Rhineland Germany. Three female mystics and a women's movement headed by a male spiritual leader offer a philosophy of interconnectedness and reverence for the earth. They include Hildegard of Bingen (1098–1179), Mechtild of Magdeburg (1210–1280), Meister Eckhart (1269–1329), and Julian of Norwich (1342–c.1415). These writers reveal a number of "ecological" themes that inspire respect and reverence for Nature and God's entire creation.

While mainstream Christian religions begin with sin, the creation-centered mystics begin with blessing. Sin is humanity's creation not God's. There is no dualistic separation between a God embodying pure goodness and a deficient, sinful creation, but all creation is itself supremely good, delightful, beautiful, and pleasureable. Each being within it is full of the divine and reveals God's goodness. God is in us and we are in God. Mechtild experienced a spiritual awakening when she saw that God was in all things and all things were in God. Julian believed that all people were enclosed within God and Hildegard wrote that "God hugs you. You are encircled by the arms of the mystery of God."

In contrast to the patriarchal religious tradition, God for these mystics, was both mother and father. Meister Eckhart imagined God lying on a maternal bed giving birth, while Julian saw the cosmos as a divine womb in which God was both Father and Mother. God is "our true Mother in whom we are endlessly carried and out of whom we will never come," she wrote. The earth too was holy, not something to be escaped from but embraced. For Hildegard the earth was a mother and living organism in whose body the seeds of all life were contained. The earth was nourished, watered, and made green by the air which was the earth's soul. Eckhart spoke of God as "a great underground river," with the earth as mediator between humans and divinity.

For the mystics, there was no dualism of body and soul as in mainstream Christianity. The body was not an enemy to be despised, but an ally to be celebrated. Eckhart believed that the soul loved the body. It was the soil in which the divine seed was planted. Mechtild admonished that the body was not something to be disdained but a safe haven for the soul. Julian believed in a soul so large that it was an endless world with God in the center. For her, human sensuality was grounded in Nature, in compassion, and in grace.

Christ was a cosmic Christ, bringer of justice, the Holy Spirit an outpouring of compassion from God and Christ. Compassion was humanity's origin, destiny, and source of justice. Making justice by way of compassionate healing was to return the Creator's gifts. Appreciating the thought of mystics such as Hildegard, Mechtild, Eckhart, and Julian, Fox argues, can help to bring an ecological awareness to the Christian tradition. Spiritual ecology is an awareness of the interconnectedness of the whole cosmos, a reverence for the earth, and compassion for all of creation.[24]

Fox's institute offers courses on cosmology and spiritual practice that bring together people of diverse religions and professions. While supported by his own Dominican Order, in 1988 he was silenced for one year by the Vatican. The grounds for his silencing were based on his references to God as "Mother," denial of the centrality of original sin, and his "fervent" feminism. Fox in turn defended his ideas by reference to the Bible and the Church's own traditions.[25]

ECOLOGICAL PROCESS THEOLOGY

Is Biblical thought ecological? Is the ecological movement religious? Is there an environmentally sensitive form of Christianity? These questions are asked and answered by process philosophers seeking a postmodern ecological worldview. For inspiration and spiritual guidance, they argue, one need not turn to the wisdom of native peoples or to eastern philosophy, but a meaningful ethic may be found within alternative western philosophies.

Ecological process theology has been developed by John Cobb of the Center for Process Studies at Claremont College and David Ray Griffin, founder of the Center for the Study of the Postmodern World in Santa Barbara, California and several of their colleagues and students. Cobb and Griffin argue that mainstream Christianity is not ecological and for the most part the current ecology movement is not Christian. As Christian theologians, they have rejected both premodern and modern forms of Christian faith. They call for a new postmodern ecological worldview that will supersede the mechanistic, dualistic, positivist worldview of the modern era. The ecological movement is the bearer of this emerging worldview.

Process philosophy owes its origins to British philosopher Alfred North Whitehead, who taught at Harvard University and to Charles Hartshorne, a teacher of Cobb at the University of Chicago. It asserts that "process is fundamental. It does not assert that everything is in process . . . but to be *actual* is to be a process." It challenges the mechanistic idea that an atom or molecule remains fundamentally the same regardless of its relations. Instead atoms acquire diverse properties in diverse relationships (or contexts). Atoms acquire different properties in different molecular arrangements because the new structures are new environments. Process philosophy thus substitutes an "ecological" theory of internal relations in which entities are qualitatively changed in interactions for the billiard ball model in which entities are like machines—independent and unchanged, affecting each other only through external relations. Atoms and molecules therefore should be viewed not as machines, but as ecosystems.[26]

Process theology holds that God created the world out of chaos (rather than *ex nihilo*) and that each stage in the evolutionary process represents an increase in divine goodness. Each *individual* thing, whether a living organism or an atom, has intrinsic value and there is a continuity between human and nonhuman experience. One's attitude toward a dog, which is a compound individual, differs from that toward a plant, which is also a compound individual but has no center of enjoyment, and toward a rock, which, as a mere aggregate, has no intrinsic value. All three, however, have instrumental value in supporting each other in the ecosystem.[27]

Process thought is consistent with an ecological attitude in two senses: (1) its proponents recognize the "interconnections among things, specifically between organisms and their total environments," and (2) it implies "respect or even reverence for, and perhaps a feeling of kinship with, the other creatures." Cobb and Griffin argue that process philosophy implies an ecological ethic and a policy of social justice and ecological sustainability:

> The whole of nature participates in us and we in it. We are diminished not only by the misery of the Indian peasant but also by the slaughter of whales and porpoises, and . . . the 'harvesting' of the giant redwoods. We are diminished still more when the imposition of temperate-zone technology onto tropical agriculture turns grasslands into deserts that will support neither human nor animal life.[28]

For Cobb's former student Jay McDaniel, intrinsic value includes the entire physical world. Atoms as individual things have intrinsic value. Rocks express the energy inherent within their atoms. They too have intensity and intrinsic value, albeit less than that of living organisms. Outer form is an expression of inner energy. The assumption that rocks have intrinsic value, however, does not mean that rocks and sentient beings would necessarily have equal ethical value, but rather that they would all be treated with reverence. This could result in a new attitude by Christians toward the natural world, one that involves both objectivity and empathy.[29]

Philosopher Susan Armstrong-Buck also sees Whitehead's philosophy as providing an adequate foundation for an environmental ethic because intrinsic value is assigned to nonhuman nature. Process is the continuity of occasions or events that are internally related—each present occasion is an integration of all past occasions. Occasions, Whitehead wrote, are "drops of experience, complex, and interdependent." The world is itself a process of fluent energy; actual entities are self-organizing wholes. Differences exist in the actual occasions that constitute each entity. Intrinsic value is not based on an extension of self-interest to the rest of nature, but on the significance of each occasion and its entire interdependent past history. Assigning preferences to biosystems is based on the degree of diversity, stability, freedom of adaptation, and integration of actual occasions inherent in each system.[30]

CONCLUSION

The main project of spiritual ecology is to effect a transformation of values that in turn leads to action to heal the planet. Whatever religion or form of spirituality one practices, it is possible to find a connection to the earth and to the political work that needs to be done to change the present way of managing resources. Some religions are more radical than others and some envision a more radical political transformation than others. With most individuals practicing some form of religion and with increasing attention to the ecological consequences of current ways of doing business, a spiritual revolution may help to support human and ecological justice in the twenty-first century. Yet skeptics argue that neither deep nor spiritual ecology goes far enough. Only through an economic transformation of the type advocated by the social and socialist ecologists of the following chapter can true ecological justice be attained.

FURTHER READING

Anderson, William. *Green Man: The Archetype of our Oneness with the Earth.*

Allen, Paula Gunn. *The Sacred Hoop: Recovering the Feminine in American Indian Traditions.* Boston: Beacon Press, 1984.

———. *Spider Woman's Granddaughters: Traditional Tales and Contemporary Writing by Native American Women.* New York: Fawcett Columbine, 1989.

Berger, Pamela. *The Goddess Obscured: The Transformation of the Grain Protectress from Goddess to Saint.* Boston: Beacon Press, 1985.

Birch, Charles, and John Cobb, Jr. *The Liberation of Life: From the Cell to the Community.* Cambridge: Cambridge University Press, 1981.

Bonifazi, Conrad. *The Soul of the World: An Account of the Inwardness of Things.* Lanham, Md.: University Press of America, 1978.

Budapest, Z. *The Holy Book of Women's Mysteries.* Oakland, Ca.: n.p., 1979.

Canan, Janine, ed. *She Rises Like the Sun: Invocations of the Goddess by Contemporary American Women Poets.* Freedom, Ca.: Crossing Press, 1989.

Cobb Jr, John and David Ray Griffin. *Process Theology.* Philadelphia: Westminster Press, 1976.

Eisler, Riane. *The Chalice and the Blade.* San Francisco: Harper and Row, 1988.

Gadon, Elinor W. *The Once and Future Goddess.* San Francisco: Harper and Row, 1989.

Gimbutas, Marija. *The Goddesses and Gods of Old Europe, 6500–3500 BC* Berkeley: University of California Press, 1982.

Hungry Wolf, Beverly. *The Ways of My Grandmothers.* New York: William Morrow, 1980.

Joranson, Philip, and Ken Butigan, eds. *Cry of the Environment: Rebuilding the Christian Creation Tradition.* Santa Fe, N. M.: Bear and Company, 1984.

Macy, Joanna Rogers. *Despair and Personal Power in the Nuclear Age.* Philadelphia: New Society Publishers, 1983.

Macy, Joanna. *World as Lover, World as Self.* Berkeley, CA.: Parallax Press, 1991.

McDaniel, Jay. *Of God and Pelicans: A Theology of Reverence for Life.* Louisville, Kt.: Westminster John Knox, 1989.

McLuhan, T. C., compiler. *Touch the Earth: A Self-Portrait of Indian Existence.* New York: Simon and Schuster, 1971.

Nelson, Richard K. *Make Prayers to the Raven: A Koyukon View of the Northern Forest.* Chicago: University of Chicago Press, 1983.

Nollman, Jim. *Spiritual Ecology: A Guide to Reconnecting with Nature.* New York: Bantam, 1990.

Orenstein, Gloria. *The Reflowering of the Goddess.* New York: Pergamon, 1990.

Seed, John, Joanna Macy, Pat Fleming, and Arne Naess. *Thinking Like a Mountain: Towards a Council of All Beings.* Philadelphia: New Society Publishers, 1988.

Sjöö, Monica, and Barbara Mor. *The Great Cosmic Mother: Rediscovering the Religion of the Earth.* San Francisco: Harper and Row, 1987.

Spretnak, Charlene, ed. *The Politics of Women's Spritiuality: Essays on the Rise of Spiritual Power Within the Feminist Movement.* Garden City, N.Y.: Doubleday, 1982.

Starhawk. *The Spiral Dance: A Rebirth of the Ancient Religion of the Great Goddess.* New York: Harper and Row, 1979.

———. *Dreaming the Dark: Magic, Sex & Politics.* Boston: Beacon Press, 1982.

Stone, Merlin. *When God Was a Woman*. New York: Harcourt Brace Jovanovich, 1976.

Vanderwerth, W. C., ed. *Indian Oratory: Famous Speeches By Noted Indian Chieftans*. Norman, Ok.: University of Oklahoma Press, 1971.

Whitehead, Alfred North. *Process and Reality*, ed. David Ray Griffin and Donald W. Sherbune. New York: Free Press, 1978.

6

SOCIAL ECOLOGY

Picture a group of eight or ten intense, younger middle-aged men and women dressed indifferently in jeans and slogan-bearing T-shirts, sitting around a cracked formica table under harsh flourescent light. . . . The remains of corned-beef-on-rye sandwiches are shoved to one side. Political posters dot the dingy walls. They are discussing the best strategy to mobilize dock workers to support a solidarity strike—to refuse to unload grapes, melons, and cherries grown in Pinochet's Chile. Through the grimy window panes, the sullen outlines of warehouses and factories are visible in the San Francisco fog.

Now picture a group—about the same size—of men and women— about the same age—gathered around a Warm Morning woodburning stove. Under the turned-up sleeves of their Pendleton shirts protrude the men's waffle-weave long underwear. . . . The women are wearing brightly patterned blouses and long skirts or sweaters and cross-country ski knickers. . . . A potluck supper of brown rice, lentil soup, and steamed vegetables simmers on the cook stove. They are discussing what crops to plant in their cooperative fields and . . . how best to present the economic advantages of organic agriculture. . . . Through the tilted panes of the passive solar herbarium, the snow-covered rolling Wisconsin fields sparkle in the mid-afternoon February sun.[1]

PROGRESSIVE ECOLOGY: "MARX MEETS MUIR"

Frances Moore Lappé, author of *Diet for a Small Planet*, and philosopher J. Baird Callicott, champion of ecologist Aldo Leopold's land ethic, set the above scenes and ask "Who could imagine, that these two groups of people could even talk to each other, much less have anything to say?" They then offer ways to unite the traditions of economist Karl Marx and preservationist John Muir.

Lappé and Callicott see the conflict between social progressives and environmentalists as stemming from seemingly antagonistic perspectives. For environmentalists, the progressive goal of the abolition of poverty and redistribution of wealth seems achievable only if nature becomes a warehouse of raw materials—a passive backdrop to industrialization. Progressives, on the other hand, view environmentalist goals of saving wilderness and improving environmental quality as benefitting white middle-class élites, while alienating the hungry, homeless, and jobless.

Yet underneath the conflict, argue Lappé and Callicott, is a common ethic of outrage over the impact of industrialization on laboring peoples and on nature. Industrial development has brought neither social justice nor a healthy environment to all people. Both the progressive and environmental movements look beyond the individual to the social and environmental whole for values by which to restructure the world. For both visions, the environment and society are the living contexts of life. Species exist in relationship to other biota and the physical environment that sustains them; humans exist as parts of an interdependent social community.

What specifically can the two movements contribute to each other? People working together can create opportunities to keep their own environments clean and remove neighborhood poverty. But a world in which there is room for both humans and wildlife cannot be achieved by biological methods or social programs alone. Expanding meaningful opportunities for employment, especially for women; food and housing subsidies; and appropriate technologies that can be repaired at the local level are methods that can help to lower population growth rates. Ecologically sensitive agriculture that helps to reduce pesticide residues and water salinization could improve social conditions. A system in which farmers have a personal relationship to their land that continues over time could maintain healthy ecological conditions. Through carefully crafted local programs, a synthesis of progressive politics and social ecology could contribute to a viable world.

Like Lappé and Callicott, many people are searching for ways to resolve the contradiction between production and ecology. Calling themselves variously social ecologists, socialist ecologists, green Marxists, and red greens, they ground their approach in an ecologically sensitive form of Marxism. Social ecologists focus on the relations of production and the hegemony of the state in reproducing those relations. Their ethic is basically homocentric, inasmuch as social justice is a primary goal, but it is an ethic informed and modified by ecological and dialectical science. The analysis of the theorists of this chapter both informs and draws on the actions of left greens, social and socialist ecofeminists, and many activists in the Third World sustainable development movement (see Chapters 7, 8, and 9).

MARX AND ENGELS ON ECOLOGY

For most people, Marxism is synonymous with the rigidity and oppression of the bureaucratic states of the Soviet Union, Eastern Europe, and China. Moreover, Marx's prediction that capitalism would generate economic and social crises that would lead to socialist revolutions in capitalist countries, led by the working classes, has not been borne out. Marx's emphasis on the lawlike characteristics of a society's economy placed less stress on the role of social movements, politics, culture, and consciousness in transforming society than on the overthrow of the mode of production. Since the 1960s, however, Marxist theorists have emphasized the processes by which people are socialized through gender, race, and class and the ways in which social movements can identify and alter those patterns. Many groups, including the New Left, democratic socialists, socialist feminists, and racial and religious minorities have found insights in the writings of Marx and Engels that promote goals of liberation, freedom, and economic equality. The same is true of ecological Marxists, who emphasize, not the control and domination of nature, but rather the ways in which ecological theories and green social movements can help to transform people's consciousness and practices toward nonhuman nature.[2]

Although Marx and Engels certainly argued that the domination of nature through science and technology would relieve humankind of the "tyranny" imposed by nature in procuring the necessities of life (food, clothing, shelter, and fuel), they were also acutely conscious of the "ecological" connections between humans and nonhuman nature. Like many critics today, they reacted against the mechanistic worldview of the seventeenth century. This mechanical materialism assumed that matter was made up of inert atoms and that all change was externally caused. Perception is explained as the result of corpuscles of light hitting an object such as a table or pencil, entering the eye, and being recorded as an impression on the brain. The individual is the passive receptor of information, just as the worker is the passive receptor of the capitalist's decision to offer minimal wages. Any worldview that casts the laborer as a powerless recipient of the ideas of a controlling élite is not healthy for her or him.

Similarly, the alternative view, prevalent in Marx's time, that the world was fundamentally spirit or idea, working itself out through history—the view of German philosopher Georg Hegel—was equally problematical. This worldview likewise rendered laborers powerless to change their destinies. What both the mechanists and the Hegelians had left out of their philosophies were social relations. People are born into a given type of society at a given time in history. Their place in that society is the perspective from which they view the world. Those in control of the society—the élite—will use the worldview to justify and maintain their hegemony. But laborers, artisans, minorities, and the poor have a choice of ways in which to view the world. They do not have to accept the mechanistic philosophy that renders them passive receptors of knowledge. More compatible with their social needs is a worldview that makes change, rather than *stasis*, central.

In arriving at a theory of social change, Marx borrowed from both the schools he rejected. With the mechanists, he asserted the reality of the material world. Matter and its manifestations in natural resources, food, clothing, shelter, and the essentials of life were real. Yet changes in the material world were not external to it, as mechanical materialism held, but internal. With Hegel, he asserted that the process of change was dialectical. The material world is continually in a process of change. This is because every event has both positive and negative forces. Everything is also not something else. It *is* by virtue of what it *is not*. The real can be defined only through contrasts. Each thing, therefore, is also its opposite. This tension, or contradiction, between a thing and its opposite destroys both and creates something new. Being (the thesis) inherently contains its own contradictions, not-being (the antithesis), and the tensions between them are a new becoming (the synthesis).

Through this dialectical process, humans make their own history. The élite society of Greece which developed philosophy and democracy, did so only because of its simultaneous dependence on slavery and sexism. The contradictions between freedom and unfreedom, between élite domination and dependency on the dominated eventually led to the downfall of the ancient social system. Medieval feudalism contained a similar contradiction between free lord and unfree serf, yet serfs, unlike slaves, had certain rights to natural resources and the manor commons. Without the serf to make in-kind payments of food and fuel, the lord by definition would not be lord. Similarly, capitalists depend on wage laborers and vice versa, but the mutual contradictions between their interests create tensions that lead to social transformation. Today, the economic dependencies of the First World on the natural resources and labor of the Third World create similar patterns of dominance. As dominators, we are ourselves dominated because of our dependence on the dominated.

Seeing the world as fundamentally process and change, however, has implications not only for society, but also for nature. Marx, in his *Economic and Philosophical Manuscripts of 1844*, recognized the interdependence of humans and nature, an idea now central to the ecological vision. People, he asserted, were active natural beings who were corporeal and sensuous and who, like animals and plants, were limited and conditioned by things outside themselves. They were both different from these objects and yet dependent on them. "The sun is the object of the plant—an indispensable object to it, confirming its life—just as the plant is the object of the sun, being an expression of the life-awakening power of the sun." Like today's ecologists, Marx recognized the essential linkages between the materials that make up the human body and nonhuman nature. "Nature is man's inorganic body," he wrote. "Man lives on nature—means that nature is his body, with which he must remain in continuous interchange if he is not to die. That man's physical and spiritual life is linked to nature means simply that nature is linked to itself, for man is a part of nature."[3]

Humans, however, differed from other animals in the way in which they obtained the essential food and energy to continue living. What distinguished humans, thought Marx and Engels, was their capacity to produce, using tools and words. The tools of animals were, in most cases, parts of their bodies, with inconsequential effects on nature. Humans, by contrast, transformed external nature with instruments that were socially organized. In different periods in history, humans organized their instruments and labor into different modes of production. Gathering-hunting, horticulture, feudalism, capitalism, and socialism are different modes of production that transform nature in different ways.

Essential to the "ecological" vision of Marx and Engels is their study of the history of human interactions with nature. Early societies, they argued, had a different relationship to nature than do capitalist societies. While pastoral societies wander, taking from nature that which is necessary for life, horticultural societies settle down and appropriate the earth's resources for their own sustenance. Humans thus modify external nature, using the local climate, topography, and flora and fauna for their own purposes. The settled community uses the earth as "a great workshop," for its labor. Human labor, on the one hand, and the earth, with its soils, waters, and organic life as instrument of labor, on the other hand, are both necessary for the reproduction of human life. Humans, isolated from society, would live off the earth as do other animals.

For the earth to be appropriated as property humans must settle on the land and occupy it. Under capitalism, the earth is bought and sold as private property. Here, according to Engels, the earth is peddled for profit. "To make the earth an object of huckstering," he wrote, "—the earth which is our one and all, the first condition of our existence—was the last step toward making oneself an object of huckstering." It is the ultimate in alienation. In the capitalist appropriation of the earth for profit, raw materials, to be taken from the earth, such as coal, oil, stone, and minerals, are the result of natural forces. They are the "free gift of Nature to capital." Nature produces them and the capitalist pays the laborer to transform them. Similarly, physical forces, such as water, steam, and electricity cost nothing. Science, likewise, costs capital nothing, but is exploited by it in the same manner as is labor.[4]

But these modes of transforming nature have unforeseen side effects. Like modern ecology, which is premised on the concept that everything affects everything else, Engels noted in his *Dialectics of Nature* that "in nature nothing takes place in isolation. Everything affects every other thing and vice versa, and it is mostly because this all-sided motion and interaction is forgotten that our natural scientists are prevented from clearly seeing the simplest things."

Engels warned that people should not boast about their ability to master nature because there were always harmful consequences of such conquests. Goats grazing on Greek hillsides prevented forests from regenerating themselves. Sailors arriving on Greek islands introduced goats and pigs that destroyed native vegetation and prepared the way for cultivated crops and weeds that obliterated native species and even the wild ancestors of grains. In Mesopotamia, Greece, and Asia Minor, those who cut down forests to plant crops did not predict that they were simultaneously destroying the collectors of moisture on which the land depended. The Italians who cut down fir forests in the Alps did not realize that they were destroying the watersheds on which the dairy industry they were introducing depended and at the same time creating the conditions for flooding the plains below. When the potato was introduced into Europe from the New World, those who did so did not consider the possibility that they were simultaneously spreading the disease of scrofula. Spanish planters in Cuba, who burned forests on steep mountainous slopes for one generation's worth of coffee profits, did not care about the erosion and ruined soil that took its toll on those that followed.[5]

"Thus at every step," Engels admonished, "we are reminded that we by no means rule over nature like a conqueror over a foreign people, like someone standing outside nature—but that we, with flesh, blood, and brain, belong to nature, and exist in its midst, and that all our mastery of it consists in the fact that we have the advantage over all other creatures of being able to know and correctly apply its laws." The more one understands the laws of nature and the consequences of human actions, he went on, the more humans will come to "know themselves to be one with nature," and that there is no inherent "contradiction between mind and matter, man and nature, soul and body." These dualisms originated in the philosophy of ancient Greece, were reinforced by Christianity in the Middle Ages, and codified by the philosophers and scientists of the seventeenth century. Their dissolution is one of the goals of the radical ecological and ecofeminist movements today.[6]

In *Capital*, Marx analyzed some of the "ecological" side-effects of the capitalist mode of production. He argued that capitalist agriculture, much more than communal farming, wastes and exploits the soil. In agriculture geared toward production for profit, the soil's vitality deteriorates because the competitiveness of the market fails to allow the large-scale owner or tenant farmer to introduce the additional labor or expense needed to maintain its fertility. The agricultural population declines as the industrial–urban population mounts, and as Marx noted (following nineteenth century chemist Justus Liebig), the marketed produce carries with it the molecules of soil–building nutrients. Large-scale agriculture and large-scale industry mutually support the enervation of both laborer and soil, breaking "the coherence of social interchange prescribed by the natural laws of life."

Capitalist agriculture, Marx observed, is progress in "the art, not only of robbing the laborer, but of robbing the soil; all progress in increasing the fertility of the soil for a given time, is a progress towards ruining the lasting sources of that fertility." It progresses only "by sapping the original sources of all wealth—the soil and the laborer." Small farming is not feasible because there is insufficient land for all to be rural land holders. Moreover, the labor of the small farmer is isolated from the larger society. Under communal production, however, there is the possibility of "conscious rational cultivation of the soil as eternal communal property, an inalienable condition for the existence and reproduction of a chain of successive generations of the human race."[7]

Industrialization, according to Marx, resulted in similar "ecological" problems. Wastes from industry and human consumption accumulated in the environment and were not reused by the capitalist unless the price of raw materials soared. Marx gave numerous examples of capitalist pollution: chemical by-products from industrial production; iron filings from the machine tool industry; flax, silk, wool, and cotton wastes in the clothing industry; rags and discarded clothing from consumers; and the contamination of London's River Thames with human waste. Yet this waste that clogged and polluted waterways was very valuable and had the potential to be recycled by industry. The chemical industry could reuse its own waste as well as that of other industries, converting it into useful products such as dyes and rugs. The clothing industry could improve its use of the waste through more efficient machinery. Human waste could be treated and used to build soil fertility. An "economy of the prevention of waste" that reused all waste to the maximum was required.[8]

Marx assumed a two–levelled structure of society: the economic base or mode of production (which consisted of the forces and relations of production) and the legal–political superstructure (Figure 6.1). Together these constituted the social formation. Different modes of production, such as primitive communism, ancient, asiatic, feudal, capital-

FIGURE 6.1
MARXIST FRAMEWORK OF SOCIAL ANALYSIS

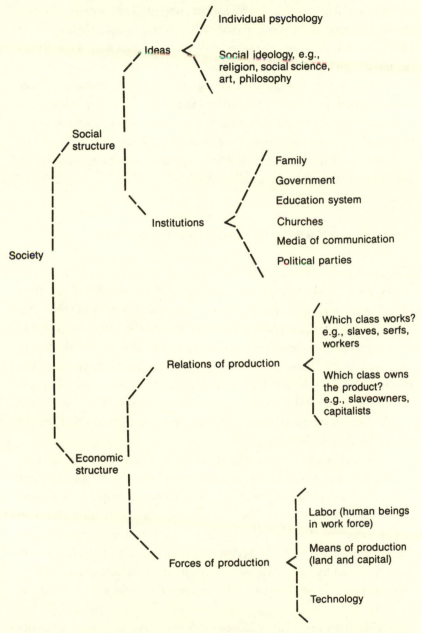

Source: Howard Sherman, *Foundations of Radical Political Economy* (Armonk, N.Y.: M.E. Sharpe, 1987) p. 44, reprinted by permission.

ist, and socialist, had different legitimating superstructures. Marx's theory of social change was based on a conflict between the material forces of production and the social relations of production. This dialectic initiates an era of social revolution in which the economic foundation breaks down leading to a change in the superstructure. Today social ecologists envision a transformation of the global capitalist economy and its legitimating mechanistic worldview to a sustainable economy and a process-oriented ecologically-based science. It would be brought about by social movements, especially those concerned with environmental health and quality of life.

ANARCHIST SOCIAL ECOLOGY

Current theories of social ecology draw on Marx and Engels' approach to "ecology" and society. Additionally, social ecologists draw their ideas from premodern tribal societies, eastern cultures, and from analyzing the ecological problems of capitalist, socialist, and Third World countries. For anarchist philosopher Murray Bookchin, social ecology is rooted in the balance of nature, process, diversity, spontaneity, freedom, and wholeness. His ideal society would eliminate all hierarchies in ecology and in society. The ecological society of the future would reclaim the fundamental organic non-hierarchical relationships of preliterate peoples.

Early preliterate societies, he argues, were organic. Although there were differences based on age, gender and kinship, such societies saw themselves as neither superior to nor inferior to nonhuman nature. They were within nature. Their differences with each other and with nature constituted what Hegel called a "unity of differences," or a "unity of diversity." Male decision-making roles in the civil sphere were balanced by the power of women in the domestic sphere. Women's central role in foraging and horticulture offset men's role in hunting.

With the continuing influence of elders, however, male authority and prestige increased and organic society broke down. Hierarchy destroyed the original balance; males became dominant over females

and children. Scarcity and warfare escalated the problems created by
the twin pillars of dominance and hierarchy, and non-egalitarian culture continued in all subsequent societies. Today dominance and hierarchy permeate all aspects of life, especially in the dominance of the
intellectual over the physical, work over pleasure, and mental control
over sensuous body. A major goal of social ecology is to abolish these
dualisms.

In an ecological society, Bookchin argues, dominance and hierarchy would be replaced by equality and freedom. An "ecology of
freedom" would reunite humans with nature and humans with humans. This would be achieved through an organic, process-oriented
dialectic that would reclaim the outlook of preliterate peoples. The
merging of their ecological sensibility with the analytical approach of
western culture would produce a new consciousness. Thus the advances of science and technology could be retained and infused with
an ecological way of living in the world. This recognizes the mutual
dependence of humans and nonhuman nature. The ecology of freedom
is rooted in a concept of ecological wholeness that is more than the
sum of its parts. "Unity in diversity" means the unfolding of the
processes of life. Bud is replaced by flower and flower by fruit, as
moments in an emerging unity. Spontaneity is the continual striving
of nature toward change and of humans toward greater self-awareness
and freedom.

Bookchin distinguishes between ecology and environmentalism.
Environmentalism adopts the mechanistic, instrumental outlook of the
modern world that sees nature as resource for humans and humans as
resources for the economy. Nature consists of passive resource objects
in habitats constructed for human benefit. Environmentalism does not
question the *status quo*, but facilitates the domination of humans over
nature and humans over other humans. Ecology, premised on interactions among the living and nonliving, contains the potential for an
alternative. Social ecology incorporates humans and their interdepen-
dences with nonhuman nature. Bookchin uses the term ecosystem to
mean "a fairly demarcatable animal-plant community and the abiotic
or nonliving factors needed to sustain it." Extended to society, it
becomes "a distinct human and natural community, [including] the

social as well as organic factors that interrelate with each other to provide the basis for an ecologically rounded and balanced community."[9]

Social ecology studies the patterns that make up the natural/social community, attempting to discern its history and inner logic. It uncovers the rich variety and diversity that are present in the community's evolution. An ecological approach to the community leaves room for spontaneity, both in nature and human nature. Biological and evolutionary forces that have resulted in the diversity found in nature must be fostered rather than controlled. Management should be like steering a ship by knowing the direction and strength of the current, waves, and winds, rather than a total domination oriented toward human benefit.

An ecological perspective challenges hierarchy in nature. An ecosystem is a food web, not a food pyramid with humans at the top. Each species is equal to every other species and to the abiotic elements that keep its cycles of life and death and predators and prey in motion. A process of development takes place in nature, "the result of an immanent dialectic within phenomena." Thus human communities and natural ecosystems interact with each other as they evolve. Not only do humans transform nature, but nature also transforms humans. Humans are the result of an evolutionary past that includes a primate and animal ancestry as well as a social ancestry. Social evolution took place within specific ecosystems. Nature is not just the passive receptor of human action, but the active transformer of human labor. Thus "nature interacts with humanity to yield the actualization of their common potentialities in the natural and social worlds."[10]

What does all this mean for the future? The world may continue down its present path toward destruction. Or, on the contrary, a reconstruction is possible in which humanity can transform its relationship to the natural world. "Our world," Bookchin believes, "will either undergo revolutionary changes, so far-reaching in character that humanity will totally transform its social relations and its very conception of life, or it will suffer an apocalypse that may well end humanity's tenure on the planet."[11]

To avoid the ultimate ecological collapse, Bookchin argues, humans must recognize and live within the requirements of bioregions. The ecosystems within bioregions limit the range of human options to control nature. Technologies, agricultural practices, and community sizes appropriate to the specific conditions of the bioregion are needed. Sufficient decentralization to avoid pollution and yet maintain and restore the region's native plant and animal life, along with new social institutions compatible with an ecological sensibility are also necessary. Diversity within the bioregion must be encouraged to reverse present trends toward crop monocultures, urban concrete, and mass culture, wiping out eons of evolution overnight. In confronting the stark possibility of the end of diversity, humans must open their imaginations to utopian visions.

Social ecology has a deep commitment not only to reversing the domination of nature, but also to removing social domination. Hierarchical and class inequalities have resulted in homelessness, poverty, racial oppression, and sexism. Of particular concern are forced and insensitive methods of controlling populations, rather than restructuring and redistributing food, clothing, and shelter.

Bookchin argues that certain deep ecologists (see Chapter 4) are insufficiently sensitive to social issues, especially regarding population, race, class, and sex. This includes some, although by no means all, supporters of Earth First!, the spiritual Greens, some bioregionalists, and some spiritual ecofeminists. To speak of a global population problem as threatening wilderness and the entire biosphere is incorrectly to analyze the roots of ecological problems by disregarding the differential impact of economic growth, especially capitalist growth, on indigenous people, marginalized rural and urban people, people of color, and women.

Social ecologists decry the idea of involuntary methods of population control, the Malthusian idea that famine, disease, and war are positive checks on population expansion, and the policy that immigration of southern and eastern hemisphere people into northern countries should be tightly restricted. Instead they support an ecologically-based development policy that uses resources in a sustainable way while raising the quality of life and redistributing the means of fulfilling basic needs.

The debate between deep and social ecologists highlights differences of opinion on where to place the core of the analysis as well as approaches to solutions. Social ecologists tend to see the problem as rooted in the dialectic between society (especially economies) and ecology, whereas deep ecologists focus on the conflict between the ecological and mechanistic worldviews. Similarly, for social ecologists, action must be focused on ecodevelopment and social justice as opposed to the deep ecologists' goal of transforming the worldview and reclaiming spiritual connections to the earth.

SOCIALIST ECOLOGY

Another alternative rooted in the Marxist tradition is socialist ecology. Socialist ecology offers an eco-economic analysis of the interaction between capital and nature and the transition to a post-capitalist society. Instead of Bookchin's emphasis on hierarchy and domination, a utopian anarchist society modelled on "nature," and a Hegelian dialectic, it envisions an economic transformation to ecological socialism, initiated by new green social movements.

Socialist ecology is spearheaded by economist James O'Connor, author of *The Fiscal Crisis of the State* and other books on economic crises. Rooted in Marx's conceptual framework, it nevertheless goes beyond Marxism to incorporate concepts of ecological science, the social construction of "nature," and the autonomy of nature. It argues that the environment and ecology are the key issues for the late-twentieth and twenty-first centuries, as evidenced by the global ecological crisis and the rapid growth of green social movements, ecofeminism, working-class anti-toxics crusades, and farm-worker anti-pesticide coalitions. It encourages an analysis of the dialectics between economy and ecology and between nature and history. Additionally, it offers a critique of existing socialist societies which have failed to address the ecological crisis and fosters thought about a reconstructive ecological socialism. In addressing the general problem of capitalism, nature, and socialism, it encourages dialogue among Marxists, Marxist-feminists, ecological Marxists, post-Marxists, left-greens, red-greens, and others.

O'Connor's theory of capital and nature is grounded in the traditional Marxian dialectic between the forces of production (technologies) and the relations of production (exploitation of labor by capital). This dialectic is the first contradiction of capitalism and leads to economic crisis and the breakdown of capitalism. But O'Connor equally emphasizes a second contradiction within capitalism, that between production and the environmental conditions of production (Figure 6.2). Marx and Engels used the term conditions of production to encompass human resources (labor), natural resources, and space. In ecological Marxist theory, these conditions of production come into conflict with the forces/relations of production. This second contradiction of capitalism leads to eco-economic crisis, initiating the transition to ecological socialism.[12]

Ecology is the basis of three conditions of production. First are the external physical conditions, what Marx called the natural elements entering into capital. Examples are the health and viability of ecosystems, such as the adequacy and stability of wetlands and the quality of soils, waters, and air. Second are the personal conditions of the laborers. Examples are the health of workers, as affected by the environment. Toxics and pesticides in the workplace, smoggy air and polluted water, unpleasant surroundings in the work environment, all affect the well-being of workers. Third are the social conditions of production, such as the means of communication among workers and managers.

In traditional Marxist theory, the first contradiction of capitalism leads to overproduction of goods. There is a decreased demand among consumers for the product. In ecological Marxist theory, however, the second contradiction of capitalism leads to underproduction. Capitalism creates its own barriers to growth by destroying its own environmental conditions of production. Ecologically destructive methods of agriculture, forestry, and fishing raise the costs of raw materials that lead to the underproduction of goods and the underproduction of capital. Soils are depleted, waters are polluted, workers' health fails, yields of produce, meat, wood, and textiles decline. In its hunger for profits, capital thus destroys its own ecological conditions of produc-

FIGURE 6.2
SOCIALIST ECOLOGY

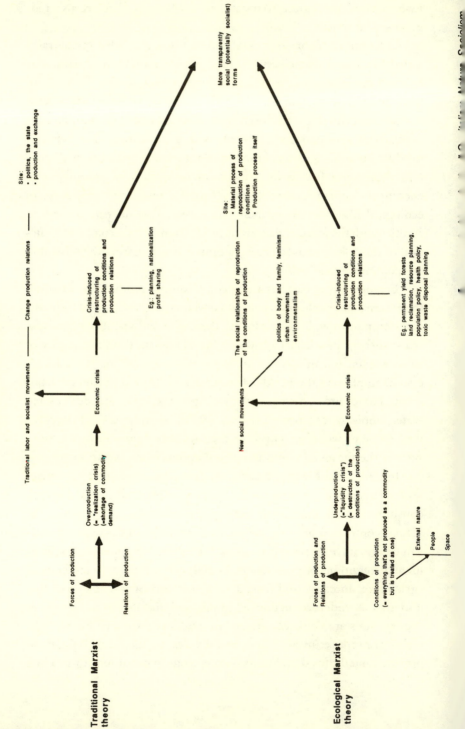

Traditional Marxist theory

Forces of production

Relations of production

Traditional labor and socialist movements ——— Change production relations ———

Economic crisis

Overproduction
(= "realization crisis")
(=shortage of commodity demand)

Crisis-induced restructuring of production conditions and production relations

Eg.: planning, nationalization profit sharing

Site:
• politics, the state
• production and exchange

More transparently social (potentially socialist) forms

Ecological Marxist theory

Forces of production and Relations of production

Conditions of production
(= everything that's not produced as a commodity but is treated as one)

External nature
People
Space

Underproduction
(="liquidity crisis")
(= destruction of the conditions of production)

Economic crisis

New social movements ——— The social relationships of reproduction of the conditions of production

politics of body and family, feminism
urban movements
environmentalism

Crisis-induced restructuring of production conditions and production relations

Eg.: permanent yield forests
land reclamation, resource planning,
population policy, health policy,
toxic waste disposal planning

Site:
• Material process of reproduction of production conditions
• Production process itself

tion. Rather than leaving Nature free and autonomous, capitalism recreates it as capitalized nature—a second nature treated as commodity and subjected to ecological abuse.

In traditional Marxism, the agencies of social transformation are the traditional labor and socialist movements that change the relations of production, through collective bargaining for example. Here economic crises make it possible to imagine the transition to socialism. In ecological Marxism, instead, the agencies of social transformation are the new ecological social movements: environmental health and safety, farmworkers' antipesticide coalitions, ecofeminist protests over groundwater toxins, leftwing green parties, and so on. Here it is ecological crises that make it possible to imagine the transition to socialism. Such crises and social movements push capitalism to respond in more transparently social and potentially socialist ways. In turn, capitalism responds by introducing more environmental and natural resources planning— sustained yield forests, environmental health policies, toxic waste disposal practices, and so on.

But in imagining the transition to an ecological socialism, socialist ecology criticizes state socialism, arguing that this is not what the new vision entails. State socialist societies have created ecological crises and fostered ecologically destructive policies, as have capitalist societies. Their planning processes nationalize production rather than democratizing and socializing it. They stifle individual creativity and are bureaucratically inflexible. They abuse and deplete nature as do capitalist societies, but do so not because of the profit-motive, but because their commitment to full employment stifles appropriate technologies and permits polllution.[13]

In an ecological socialist society, Nature will be recognized as autonomous, rather than humanized and capitalized. Ecological diversity, an ecological sensibility, and a science of survival based on the interrelatedness of living organisms and the environment will all be needed and valued.

What is an example of such an ecology of survival? One such case history is the use of biological insect controls in Nicaragua. Before the Nicaraguan revolution of 1979, agricultural production was dependent on heavy applications of pesticides to produce high cotton exports.

Broad spectrum chemicals destroyed natural insect enemies, created new chemically-resistent pests, and caused high numbers of pesticide poisonings among workers. A pesticide treadmill set in, in which a cotton export economy became dependent on increasing amounts of pesticides to maintain yields, fueling the profits of foreign chemical companies. After the overthrow of the Somoza regime, the new socialist government stepped up the use of Integrated Pest Management (IPM) techniques and revolutionized the forces of production.

Integrated Pest Management uses biological methods of controlling insect pests as its core. It depends on the careful monitoring of pest levels by trained field aides who assess when the economic threshold of pest damage has been reached, as opposed to spraying broad spectrum chemicals on predetermined calendar dates. Pesticides are applied only in limited amounts and in narrow ranges. Plants are cut and plowed under the soil between seasons to avoid carryover of pests. Before the Sandinista revolution, regional IPM programs had been difficult to implement because not all capitalist growers in an area cooperated. The restructuring of farms under the new government created new relations of production which allowed new forces of production such as IPM to take root. These new productive forces fostered better conditions of production by improving both the health of the soil and the health of the workers. The government was able better to plan production, train IPM field hands, save on the enormous costs of pesticides, and achieve higher yields.

IPM as a force of production creates independence as opposed to chemical-company dependence and creates jobs for field workers. Yet the problem of outside markets for the sale of agricultural produce was obfuscated by trade embargoes imposed by the United States and by the continuing devastation of the country by the war with the Contras. Thus, despite increasing independence in production, dependence on world markets and politics hindered economic stability.[14]

DIALECTICAL BIOLOGY

Does social ecology go beyond applied sciences, such as IPM, to include scientific method itself? In *The Dialectical Biologist*, Richard

Levins and Richard Lewontin argue that science done in the style of Marx and Engels is based on different assumptions than those of mechanistic science. Whereas mechanism is Cartesian, reductionist, and positivist (i.e. based solely on the validity of empiricism and mathematics), a dialectical perspective on science emphasizes change, historicity, and social construction. What is accepted as knowledge in any given period changes over time. What is socially and politically acceptable in any given society sets the goals and parameters of scientific investigation.

Dialectical science is based on four assumptions about the nature of reality. First, the whole is a relation among parts, rather than a sum of basic elements. These parts do not exist apart from the whole, but only in relation to it. Second, the properties take their meaning from the whole. They exist only in interaction with it. A person acquires the property of flying only in relation to a social-technological system of aluminum extraction and construction, petroleum, and pilots. Third, parts and wholes interpenetrate. Causes become effects, subjects become objects and vice versa. The environment shapes the individual and the individual shapes the environment. Both nature and people are actors in the making of history. Fourth, change is primary. It is the fundamental constant. Stability is only a momentary balance. In every object there are oppositions and contradictions that bring about change. Harmony, stability, balance, stasis, equilibrium, adaptation, and so on are illusions. Even the so-called fundmental constants of science, such as the mass of the electron and Plank's constant may change slowly over eons. If so, present assumptions about the origin and evolution of the universe could be seriously challenged.

To say that science is socially constructed is to recognize that scientists, like everyone else, bring to their work a set of assumptions about the world. While scientists try to be explicit about the mathematical and empirical assumptions and the laws that underlie their scientific papers, they are influenced by other implicit assumptions about society and the world that help to determine the kinds of research problems that are investigated and funded. The results of their research make up the theoretical basis of ongoing scientific investigations. What scientists see, hear, and attend to is influenced by a prior set of ideological beliefs. "Knowledge is socially constructed," according to Levins and Lewontin, "because our minds are socially constructed and because individual thought only becomes knowledge by a process of being accepted into social currency."[15]

Mechanistic science deals with a very small number of the possible relationships that exist in the world. It attempts to explain observable phenomena in terms of small hidden parts (atoms and molecules) as underlying causes. Dialectical science by contrast does not presuppose a hierarchy of parts and causes. How one divides up the whole depends on the particular problem. Ecology looks at interactions among parts of a community rather than setting up hierarchies among higher and lower forms. A given species such as a migrating bird or caribou may be a part of several different communities at different times.

In a dialectical worldview, objects of natural laws become subjects that may change the apparently constant laws themselves. For example, the origin of life from inanimate matter changed the law which enabled life to originate because living organisms converted the atmosphere from a reducing to an oxygen-rich atmosphere. Mechanistic science separates internal from external causes, holding one constant while changing the other. Thus the environment triggers changes in the organism, as when a living thing adapts to environmental change. Or conversely, an internal change initiates development as in the case of an embryo. A dialectical approach looks at the effects of both environment and genetic makeup together.

Dialectical science considers change as a tension between opposites. Thus in predator-prey relationships, the process of predation is propelled by the deathrate of the prey and the birthrate of the predator, and vice versa. The interaction between the two opposites causes fluctuations in population. As change occurs the initial conditions change. Levins and Lewontin maintain that contradictions in nature are not only political, but ontological, that is, fundamental to being itself. "Opposing forces lie at the basis of the evolving physical and biological world. Things change because of the actions of opposing forces on them, and things are the way they are because of the temporary balance of opposing forces." Rather than change, it is stability and equilibrium that bear explanation. Opposing processes regulate and stabilize an object, as when blood sugar rises in response to sugar ingested or falls with the release of insulin from the pancreas. Similarly, blood pressure is regulated by processes in the kidneys. In each case opposing forces mutually regulate each other to achieve homeostasis.[16]

Social ecology is criticized by deep ecologists for its ponderous and, to some, outdated Marxist theory, for its failure to offer any analysis of a transpersonal or ecological self, and for its lack of any realistic scientific alternatives based on dialectics. To spiritual ecologists, it fails to offer anything beyond the immediate fulfillment of economic and material needs and even denigrates spiritual needs. To ecofeminists, it fails to incorporate an analysis of socially constructed gender differences and is impoverished by its lack of proposals for overcoming gender/environment problems. Nor have social ecologists given adequate attention to environmental ethics. They have not shown how a basically homocentric ethic oriented toward social justice can also be sensitively informed by ecological principles. The debates among these various camps of radical ecologists, however, are important, as they push each other to rethink and reevaluate their own proposals for change.

CONCLUSION

Social ecology emphasizes the human implications of systems of economic production on the environment. Both capitalism and state socialism produce externalities that disrupt nature. Social ecology envisions a world in which basic human needs are fulfilled through an economic restructuring that is environmentally sustainable. While social ecologists would like to see world population stabilize at a level that is compatible with environmental sustainability, they deplore any programs that result in genocide, racism, or callous disregard for human rights in bringing about a demographic slowdown. Instead, economic programs that provide for basic needs, healthcare, security in old age, and employment are the pathways that will bring about a demographic transition in developing countries and equalize the quality of life in both developed and developing countries. Finally, social ecology advocates a science oriented toward social values and the recognition of change, rather than stability, as the basic premise on which to understand the natural world. It is similar to deep ecology in calling for a major transformation in worldviews and a process-oriented science, but differs from it in its emphasis on the human condition, the economic basis of transformation, and a homocentric as opposed to an ecocentric ethic. The ideas of deep, spiritual, and socialist ecologists find expression through the movements discussed in Part III.

FURTHER READING

Bookchin, Murray. *Post-Scarcity Anarchism*. San Francisco: Ramparts Press, 1971.

———. *The Ecology of Freedom*. Palo Alto, Ca.: Cheshire Books, 1982.

———. *The Modern Crisis*. Philadelphia, Pa.: New Society Publishers. 1986.

———. *Remaking Society*. New York: Black Rose Books, 1989.

Eckersley, Robyn. *Environmentalism and Political Theory: Toward an Ecocentric Approach*. Albany, N. Y.: State University of New York Press, 1992.

Gorz, André. *Ecology as Politics*. Trans. Patsy Vigderman and Jonathan Cloud. Boston, Ma.: South End Press, 1980.

Grundmann, Reiner. *Marxism and Ecology*. New York: Oxford University Press, 1991.

Lappé, Frances Moore and Joseph Collins. *Food First: Beyond the Myth of Scarcity*. New York: Ballantine, 1979.

Levins, Richard and Richard Lewontin. *The Dialectical Biologist*. Cambridge, Ma.: Harvard University Press, 1985.

O'Connor, James. *The Corporations and the State: Essays in the Theory of Capitalism and Imperialism*. New York: Harper and Row, 1974.

———. *The Fiscal Crisis of the State*. New York: St. Martin's Press, 1973.

———. *The Meaning of Crisis*. New York: Basil Blackwell, 1987.

Parsons, Howard, ed. *Marx and Engels on Ecology*. Westport, Ct.: Greenwood Press, 1977.

Schnaiberg, Alan. *The Environment: From Surplus to Scarcity*. Oxford, Eng.: Oxford University Press, 1980.

III

MOVEMENTS

7

GREEN POLITICS

The environmental movement in the 1990s has arrived at a crossroads. At the intersection, several branches take off in different directions. The avenue on the right is newly paved and its center strip is painted white. Down this highway travel large numbers of established environmental groups, carrying banners that read "Wilderness Forever," "Save the Birds," "Clean up the Oilspill." Known as the Group of Ten, the ranks of these organizations have swelled markedly in response to the Reagan "anti-environmental" decade of the 1980s. The road they travel leads to the nation's capital where the heads of each group meet regularly to divide up issues and pledge support for each other's actions. Ensconced in sleek Washington headquarters, they vie with other lobbyists for power to influence the executive, legislative, and judicial branches of government. They breakfast with corporate leaders and bankers to work out long term environmental deals and debt for nature swaps.

At the intersection, a branch toward the left is under new construction. Still rocky and covered with multi-colored soil, the construction work is being carried out by grassroots activists. Concerned that the road pass through clean air and waters and that its workforce be treated fairly, the builders stop frequently to oppose dumpsites, pollutants, and the victimization of peoples of color. At a bend, an obscure trail turns off to the left. Down it move those bent on civil disobedience in defense of nature—Earth First!ers and animal liberationists.

At the center of the crossroads, a new road is still in the planning stage. Here people dressed in green clothing are preparing for construction by painting signs reading "We are neither left nor right, we are in front." These Greens have formed Committees of Correspondence that communicate with members around the world and they reach decisions by consensus. Their work is slow, as they must appeal to many local governments for support for their projects.

Green political groups act to change the ways in which both capitalist and socialist production are reproduced through laws and governance. The various movements try to resolve the contradiction between production and reproduction by bringing pressure to bear on specific aspects of social reproduction. The older established environmental organizations in most countries concentrate their energies on government, pressing for stronger environmental laws, lobbying for specific bills, and challenging existing laws through the courts. The Greens, however, wish to transform politics itself in new, more truly democratic, and especially green directions (see Figure I.1).

Grassroots activists focus on the ways in which daily life is reproduced in neighborhoods and local communities, demonstrating loudly for clean water, air, and healthy food, and against toxic and nuclear threats to biological reproduction. Their strategies include marches, demonstrations, and negotiations at points of production, such as corporation headquarters, incinerator sites, and toxic dumps. Other

groups, such as Earth First!ers, Greenpeace activists, and animal liberationists, pinpoint threats to the reproduction of non-human nature
(wolves, whales, rainforests, and wilderness), resulting from the production of marketable commodities. Their strategies include confrontations at points of resource extraction, such as lumbering, whaling,
fishing, and agribusiness locations.

What are the goals and projects of these builders of roads to the
twenty-first century? What are their tactics for changing society? What
are their ethical frameworks? Who are their members?

THE GROUP OF TEN

The Big Ten environmental organizations have their origins in the first
and second waves of environmentalism. The oldest groups, such as
the Sierra Club and the National Audubon Society, originated in the
late nineteeth century and made their first national impacts during the
progressive conservation movement of the early twentieth century.
The progressive movement was initially an effort to conserve forests,
rangelands, and parks for the benefit of middle-class America. Its ethic
was homocentric, rooted in Gifford Pinchot's utilitarian maxim of "the
greatest good, for the greatest number, for the longest time." John
Muir's Sierra Club emphasized wilderness preservation, but its effort
to create national parks was supported by the railroads which reaped
profits from middle-class tourism, while the Sierra Club itself was
divided on such issues as the need for dams for city water supplies
versus wilderness recreation (see Chapter 3). More recent environmental organizations, such as Friends of the Earth and the Natural Resources Defense Council, are children of the environmental movement
of the 1970s and are likewise supported by a middle-class constituency.

The Group of Ten includes the following organizations:

Environmental Defense Fund
Environmental Policy Institute
Friends of the Earth
Izaak Walton League of America

National Audubon Society
National Parks and Conservation Association
National Wildlife Federation
Natural Resources Defense Council
Sierra Club
Wilderness Society

The Big Ten have traditionally focused much of their attention on the legislative process, calling on their constituencies to support particular bills and lobbying efforts, and on the judicial process, challenging legislation and executive decisions that fail to meet high environmental standards. Each organization focuses on particular types of issues, with one taking the lead on a given problem, usually supported by the others. All increased their numbers during the 1980s in response to perceived cutbacks in governmental support for environmental protection. They draw their members and staffs primarily from white middle-class educated Americans concerned with issues of wilderness and wildlife preservation in the United States and the Third World. The Sierra Club increased from 80,000 to 500,000 during the 1980s, the National Wildlife Federation added 8000 new members a month, and the Natural Resources Defense Council doubled its numbers between 1985 and 1989.

Big Ten budgets show increased contributions from corporations and their boards include corporate executives (see Table 7.1). Action strategies are designed to retain the support of wealthy supporters. The National Audubon Society, for example, regularly sends out slick catalogues of consumer birdwatching items designed to appeal to wealthy donors. The National Wildlife Federation added a Corporate Conservation Council in 1982 dedicated to economic progress through resource conservation. The head of Waste Management Inc. (charged with numerous environmental violations), is a board member. Only 22 percent of NWF's budget comes from members, with another 15 percent from magazine subscriptions for school children. Corporate donors include Amoco, ARCO, Coca Cola, Dow, DuPont, Exxon, General Electric, General Motors, IBM, Mobil, Monsanto, Tenneco, Westinghouse, and Weyerhauser.

According to former Environmental Protection Agency chief,

Table 7.1

Environmental Organizations and Funding Sources, 1988 Budgets

National Wildlife Federation
  $63 million; Members 22%; Magazine Subscriptions 15%; Corporate donors: Amoco, ARCO, CocaCola, Dow, Duke Power, DuPont, Exxon, GE, GM, IBM, Mobil, Monsanto, Tenneco, USX (formerly U.S. Steel), Waste Management, Westinghouse, Weyerhauser. Matching grants from Boeing, Chemical Bank, Citibank, Pepsi, the Rockeferrer Group, United Technologies.

Audubon Society
  $38 million; Members $10 million. Corporate donors include the Rockefeller Brothers Fund, Waste Management Inc., General Electric, GTE, Amoco, Chevron, Dupont, Morgan Guaranty Trust. Donations under $5000 from Dow Chemical, Exxon, Ford, IBM, Coca Cola.

Sierra Club
  $19 million, Members 64%. Corporate donors matching gifts programs through which companies match employee contributions include funds from ARCO, British Petroleum, Chemical Bank, Morgan Guaranty Trust, Pepsi, Transamerica, United Technologies, Wells Fargo.

Natural Resources Defense Council
  $11 million; Members 40%

Wilderness Society
  $9 million; Members 50%; Corporate donors include Morgan Guaranty Trust and Waste Management.

Source: Brian Tokar, "Marketing the Environment," *Zeta Magazine* (February 1990), pp. 16–17, compiled from 1988 Annual Reports of major environmental organizations.

William Ruckelshaus, "the strongest supporters of a forceful EPA are the industries it regulates. They want government to set reasonable standards and they want the public to know they are being enforced."[1]

During the 1980s, mainstream environmentalism moved further from grassroots confrontation and closer to corporate cooperation. The growing sense that the Big Ten were intimately connected with reform environmentalism led to disenchantment among those who wanted to use direct action to assert the rights of women and minorities, to protest corporate pollution, and to save wilderness and endangered species. These groups, from housewives confronting neighborhood waste spills and minorities protesting community incinerators to Greenpeace sailors saving whales and Earth First!ers sitting in redwoods, used marches and demonstrations, camp-ins and sit-ins, petition drives, civil disobedience, and street theater to publicize their issues.

POPULISM

During the 1980s, grassroots activists took on local hometown, backyard issues. Deeply skeptical of the assurances of government, industry, and mainstream environmentalism, much of the populist movement has centered on issues of human health—toxic chemicals in water, air, and soil to rising cancer rates and reproductive harm. The ethic of anti-toxic coalitions is homocentric, but there is a recognition that what is healthy for humans is also healthy for other species and the environment as a whole. Many of the most vociferous proponents are women (see Chapter 8). Love Canal activist Lois Gibbs's Citizen's Clearing House for Hazardous Waste began recording the uprising against toxics in 1983 through its newsletter, *Everyone's Backyard.* Annual grassroots conventions, held since 1986, link local campaigns by bringing out issues of social justice. The National Toxics Campaign, which has published *Toxic Times* since 1988, brings together activists and their movements around the world. The majority on its board of directors are women.

Over 200,000 hazardous waste sites exist at present in the United States and the list continues to grow. Cleaning them up under the Superfund law is only one part of a program to deal with the effects of toxic chemicals. A toxics prevention campaign is also essential to curtail the production of dangerous chemicals when safer alternatives exist. State legislative and initiative campaigns and a national policy on hazardous wastes sponsored by activists are a high priority.

The anti-toxics movement opposes toxic waste incinerators that, if constructed as proposed, could release millions of pounds of chemicals into the atmosphere by the year 2000, creating "local sacrifice zones." The cure makes the illness worse. Incineration of wastes increases acid-rain forming gases and toxic ashes. The resulting carbon dioxide contributes to global warming, while chlorinated chemicals add to ozone depletion. Incinerators, contends the National Toxics Campaign, violate the Resource Conservation and Recovery Act which mandates waste reduction as the strategy for controlling waste.[2]

Instead, waste production should be reduced at the source by the industries themselves. This entails: (1) reducing and eliminating waste output, (2) recycling, reusing, and exchanging waste that cannot by eliminated, (3) ceasing production of unneeded products that contribute to waste, (4) treating and destroying nonrecyclable waste at the production site rather than releasing it into the environment.

To protect people victimized by toxics, the National Toxics Campaign advocates a Bill of Citizens' Rights that includes:

The right to be safe from harmful exposure.
The right to know.
The right to cleanup.
The right to participate.
The right to compensation.
The right to prevention.
The right to protection and enforcement.

The grassroots movement against toxics advises local groups to plan a clear strategy for action by asking the questions: What do we want? Who can give it to us? How do we make them do it? The 1986 federal Community Right to Know law allows citizens to obtain information on chemical emissions affecting their neighborhoods. They can then work with local industries to clean up chemical waste production. Groups should organize so as to include all people directly affected, not just property owners. They should promote participation and unity among all ethnic groups which have a stake in the outcome.[3]

MINORITY ACTIVISTS

November 12, 1988. A multi-racial crowd of one thousand women and men marches down a street in East Los Angeles. Chanting and waving banners proclaiming, "El Pueblo Parará el incinerador" ("The People Will Stop the Incinerator,") and "Pueblo que lucha, triunfa!" ("People who struggle, win!"), they arrive at the site of a proposed toxic waste incinerator. Sixty year old Aurora Castillo of Mothers of East Los Angeles (MELA) seizes the microphone. "They thought the people were a sleeping giant. We're not sleeping anymore." Assemblywoman Lucille Roybal-Allard follows, "They think that if they pick a poor community, they won't have any resistance. We are here to prove they are wrong."[4]

The two women are members of a minority-led coalition that sucessfully brought together neighborhood, environmental, and radical activists to halt a Los Angeles plan to construct a waste incinerator in an inner city neighborhood. With the support of minority lawmakers and grassroots volunteers, the neighborhood built a coalition that tapped local opposition, while middle-class environmental and slow-growth groups supplied expertise and labor. Although mainstream environmental groups held back, Greenpeace sent people to help organize the campaign and experts to testify before the city council.[5]

Grassroots campaigns in urban environments are increasingly initiated by people of color and respond to issues facing low income people. The movement took off in 1982 when a Warren County, North Carolina neighborhood that was 60 percent African American and 4 percent Native American staged an unsuccessful uprising against a proposed PCB (polychlorinated biphenyls) disposal site. A 1987 United Church of Christ report on "Toxic Wastes and Race in the United States," showed that "communities with the greatest number of commercial hazardous waste facilities had the highest composition of racial and ethnic residents." Fifty-eight percent of the country's blacks and 53 percent of its Hispanics live in communities (such as Emelle Alabama, Houston Texas, and Chicago's south side) where hazardous waste dumping is uncontrolled.[6]

In 1989, a Citizens for a Better Environment study of Richmond

California documented 350 industries that used hazardous chemicals and 210 toxic chemicals that were released into neighborhoods where African American and Hispanic populations are concentrated. Fourteen hundred people assembled at the North Richmond Baptist Church in April of 1990 to hear black presidential candidate Jesse Jackson campaign for cleaner air for inner city people. "In Selma," Jackson proclaimed, "we marched for the right to vote. This morning we are gathering for the right to breathe." Jackson's message urged the churchgoers to become stewards of the earth and to negotiate with the polluters for a better environment.[7]

Inner city air and soil are contaminated with lead from chipping housepaint and auto emissions. A 1988 study conducted by the federal Agency for Toxic Substances and Disease Registry showed that black children were four times as likely to encounter lead poisoning as white children. A disproporionately large number of minority workers have blood lead levels above those considered safe.[8]

Farm-workers, 80 to 90 percent of whom are of Hispanic origin, are bombarded daily by pesticides. They live with rashes, dizziness, headaches, and nausea. Most are not informed of or instructed in pesticide safety regulations and fear loss of jobs if they protest. To underscore the problem faced both by workers and the public, the United Farmworkers Union has called for a nationwide ban on table grapes and other foods subjected to pesticide contamination.[9]

Native American tribes have been offered large sums of money for allowing their lands to be used as toxic waste dumps. The Environmental Protection Agency has held back on providing help and financial assistance to native peoples. Native groups have therefore created their own movements to preserve their rights. To combat radioactive and toxic waste dumping on native American lands, Jessie DeerInWater founded Native Americans for a Clean Environment (NACE) in 1984. Alarmed by her discovery that Sequoyah Fuels planned to inject radioactive waste into a fault line in Vian Oklahoma, she alerted her Cherokee sisters and brothers. NACE then went on to fight the conversion of radioactive and toxic waste into fertilizer, or raffinate, implicated in the discovery of a nine-legged frog where the fertilizer was applied.[10]

Such actions and statistics underscore the need for the Group of Ten and minority groups to come together on environmental and health issues. Black activist Cora Tucker, founder of Citizens for a Better America, describes the differences in priorities:

> The environmental issues are cut in such a way that Blacks and Hispanics don't feel like it's their issues. The traditional environmental groups talk about how we got to do something about saving the yellow-bellied sapsucker. Black people are more interested in saving their children that they see dying in their arms (from toxic wastes). It's hard for white folk to understand that we care about the environment.[11]

Carl Anthony, a black architect and co-founder of Earth Island Institute's Urban Habitat Program, observes that the environmental movement

> has tended to be racially exclusive, expressing the point of view of the middle- and upper-middle income strata of European ethnic groups in developed countries. It has reproduced within itself prevailing patterns of social relations. Until recently, there has been little concern for the environmental needs and rights of historically disadvantaged groups in developed countries. . . . Can we ignore the underclass trapped in American ghettos while claiming to speak for reconciliation of economic growth with environmental integrity? If we are to restore the cities, we must invest in the future of the people who live there.

Anthony suggests that environmentalists and inner city organizations can work together on restoration projects that promote tree planting, horticulture, urban farming, wilderness outings, and environmental education for minorities, as well as building coalitions that address issues such as toxic waste dumping.[12]

Despite the new activism of minorities the green movment is largely white. The Southern Organizing Committee for Racial Justice and the United Church of Christ's Commission for Racial Justice have faulted the Group of Ten for racism in hiring. None of the groups has minority directors or managers and their staffs have less than one percent minority representation. The national environmental groups admit that minority involvement must be increased, but also cite low salaries as a deterrant to hiring minorities. The Audubon Society is concerned that "not one major environmental or conservation organization can boast of significant Black, Hispanic, or Native American membership." A Sierra Club spokesperson believes that "the ethnic diversity of public policy is going to increase during the next century. If the environmental community does not mirror that change, our . . . ability to influence public policy makers will deteriorate." But more pragmatically mainstream environmentalists fear that their credibility on positions that affect the Third World is compromised by their largely white middle-class North American base. Yet gaps between inner city survival issues and saving the Antarctic wilderness are perhaps too wide to be bridged effectively in the immediate future.[13]

THE GREENS

At the international level, Green politics have become a major force for ecological change. Australia's United Tasmanian Group, formed in 1972, and New Zealand's Values Party, formed a few months later, were the first political parties with green platforms to challenge established parliamentary systems. The West German Greens (*die Grünen*) emerged in the early 1980s from a mass movement that used direct action to confront local community issues. They drew on people who had participated in such "basis" movements as the anti-nuclear, ecology, women's, peace, urban squatters, gay rights, Third World solidarity, and youth movements. They burst onto the international scene in 1983 when they won enough votes (5.6 percent) to be seated in the West German National Assembly (the *Bundestag*). In the elections following the German reunification of 1990, the West German Greens lost their representation, but the East Germans gained eight seats.

The political platform of the Greens is based on four pillars: (1) Ecology (2) Grassroots Democracy (3) Social Justice (4) Nonviolence. Along with the four pillars, six additional principles comprise the Ten Key Values: (5) Decentralization, (6) Community-based Economics,

(7) Postpatriarchal Principles (8) Respect for Diversity (9) Global Responsibility (10) Future Focus.[14]

Greens are divided between party and movement politics. With West German parliamentary representation a division appeared between the "realos," who held the majority of seats and adapted to the pragmatics of the parliamentary framework, and the "fundis," who held to the original values and formed the majority in the collective movement. In a 1988 manifesto, the realos stated that "the ecological threat to industrial society can be turned around only in the framework of the existing system." In an effort to appeal to the "enlightened" middle-class, they called on multinational corporations to adopt environmental standards. They opted for working with the state to solve environmental problems and advocated entering into coalitions with other parties. The realos were opposed by the movement approach of the fundis, who included left Greens, eco-socialists, and radical feminists. With the further success of the Greens following the West Berlin elections of 1989, the realo Greens formed red-green coalitions with the Social Democrats, the traditional labor party.[15]

Similar divisions between party and movement politics appeared in other European Green parties. The Italian Greens (i Verdi), founded in Florence in 1984 from local groups of anti-nuclear, ecology, citizen, and religious activists, decided two years later to present party backed Green Lists for local, regional, provincial, and national elections. The parliamentary Verdi hold that party representation allows greater access to resources and to the centers of decision making. The French Greens (les Verts) began as the Ecologist Party in the early 1980s and in 1992 won 14 percent of the vote. Thus the established parties are eager to form alliances and to capture their votes. In a country which depends on nuclear power for most of its energy needs, the French Greens are notable for their anti-nuclear and pro-environmental positions.[16]

In Sweden the Green Party (Miljöpartiet de gröna), formed in 1981, won numerous local offices and achieved national parliamentary representation in the elections of 1988, but lost that representation in the 1991 elections. The party promotes the attainment of a balance with nature through self-sufficient organic agriculture, abolition of nuclear power, development of alternative energy sources, decentralization of living and working environments, reduction of dependence on automobiles through increased public transit, and the manufacture of products that satisfy basic human needs. The party's constitution requires that a minimum of 40 percent of each sex be represented on each of its three central committees. It also has some 30 issue-oriented committees. Women are most active on those concerned with peace, housing, schools, children, medical care, agriculture, and culture. Men are dominant on those that deal with the economy, energy, science, labor, and international issues. The other committees have a more even gender division.[17]

In Canada, Green parties were formed in 1983 and have been active in British Columbia, Ontario, and Nova Scotia. According to the Green Web of Nova Scotia, a division exists between movement and party people. The issues are the accountability of those involved in politics and control by the grassroots constituents. "Much of the best environmental work that is being done in Canada," states Green Web, "is being done by groups who are completely independent of the green parties that exist." Party Greens respond that people have a choice as to whether they want to join environmental groups or work within the party. However, "working with the Green Party means that you are committed to a broader vision of social change than strictly environmental issues as with pressure groups."[18]

In the United States, the Greens are not a party, but a movement. Local groups are coordinated through the United States Green Committees of Correspondence. In 1987 they developed the Strategy and Policy Approaches in Key Areas process, known as SPAKA. Then in 1989, some 200 green groups throughout the United States, representing hundreds of people submitted position papers that were discussed and incorporated into a Green Program USA at a nationwide meeting in Eugene, Oregon. The delegates attempted to reach consensus on each position paper. When consensus was not achieved, suggestions and blocking statements were submitted. The Program Text was then returned to local groups and the membership at large for discussion and refinement. Women constituted about half of the Eugene gathering and held key positions, but equity was still a goal rather than a reality.[19]

The incipient Green Program USA includes points of view on

such topics as social justice, peace and nonviolence, politics, general economics, water, forests, food and agriculture, native Americans/ indigenous peoples, animal liberation, life forms, eco-philosophy, ecofeminism, and green spirituality. On each topic, the Program includes philosophical principles as well as specific recommendations for actions. It favors structural changes that "promote economic democracy in production, distribution and consumption under the control of an informed and empowered public, constrained by the principle of sustainability and committed to the integrity of the Earth and its regenerative powers." Focusing on waste as a problem of production rather than disposal, it argues for a sustainable, closed loop resource economy in which the by-product of one system becomes the input and source-material for another. It promotes energy conservation and efficiency with the goal of making local communities energy self-sufficient. In agriculture it calls for an ecologically-based sustainable agriculture that would promote regional self-reliance and an end to factory farming and capital-intensive, highly-mechanized chemical agriculture.[20]

The US Greens have supported Green candidates in local elections and Green parties are beginning to form at the state level. Green activists won elections in 1989 to city councils in such towns as Gloucester, Holyoke, and Cambridge Massachusetts, Ithaca New York, and Chapel Hill North Carolina. In Alaska, which recognizes the Green party, a Green candidates won the election for mayor in the town of Cordova. In California, Greens already serve on local city councils, water borards, and planning commissions, and a Green Party will appear on the state ballot in 1992. At the other end of the spectrum an emerging Left Green Network held a 1989 organizing conference that embraced principles such as anti-capitalism, social ecology, and women's, gay, and lesbian liberation. The network calls for an independent radical politics outside the Democratic and Republican parties and promotes an ecologically-oriented cooperative commonwealth based on decentralized, democratic, public-ownership of property and guaranteed housing, healthcare, and employment.[21]

US Greens are divided in their theoretical and ethical allegiances. Many espouse deep ecology and an ecocentric ethic, which views humans and other species as integral parts of the ecological whole, as the rational foundation for their politics. A strong contingent is motivated by the need for humans to reclaim deep spiritual connections to nature and uses ritual at Green gatherings as a mode of energizing people for action (see Chapters 4 and 5). Left Greens, on the other hand, are informed by social ecology's homocentric ethic of justice for all people. This homocentric ethic, however, is not the utilitarian ethic of the Progressive conservation era, but one that is enriched by an understanding of the place of humans in the interconnected ecological world and the prior history of human domination of other peoples and nonhuman nature (see Chapter 6).

GREENS IN THE SECOND WORLD

While green politics in the First World attempt to repair ecological damage from capitalist production and to transform production itself, greens in the Second World are equally alarmed by the ecological impacts of expanding industrial production in their own state socialist societies. And as many Second World countries move toward market economies, stategies for dealing with ecological problems become increasingly complex.

An emerging green movement is taking shape in the commonwealth countries of the former Soviet Union and in eastern Europe. In the Soviet Union, Stalin and Breshnev had used the slogan "Fight Nature." Industrialization targets that had to be met every five years resulted in the rapid growth of heavy industry with attendant environmental problems. But environmental activists, encouraged by former President Gorbachev's *perestroika* (restructuring) and *glasnost* (openness) programs, have pointed to structural problems in the Soviet economic planning process, as well as to pollution from industry, nuclear power, and public works construction. In the 1970s scientists and environmentalists who objected to construction of a dike to prevent flooding in Leningrad's Neva River Delta were fired. Now they form St. Petersburg's Delta Group and speak freely. In May 1991, a green party was formed in an attempt to safeguard the environment in the transition from a planned to a market economy.

In Hungary, growing citizen concern about environmental mismanagement broke out in 1988 over the planned construction of a series of dams on the Danube River. It was followed by local protests over air pollution in Budapest, the disposal of nuclear-power plant wastes, and the effects of bauxite mining.

Polish citizens' environmental movements have followed in the wake of Solidarity's labor movement. Farmers in the beautiful mountainous region around Rabka protested the planned expropriation of their traditional land holdings for redevelopment, arguing that the region's ecology and climate would be destroyed. Their protest was joined by members of the Polish Ecological Club and the Association of Polish City Planners and succeeded in halting the development. Protecting the environment was also of significance in Gloskow, where development was opposed by market gardeners. Of even wider concern were the health effects of a proposed nuclear power plant for nearby urban populations and the destruction of natural habitat required by the construction. Long term environmental protection for the benefit of the common good took precedence over short term economic gain.

A Yugoslavian Green Union was founded in 1988. It argued for alternative ecologically sound development, rather than the bandaids traditionally applied as environmental problems emerged. One hundred people demonstrated in Slovenia over the mismanagement of the environment and natural resources. By 1989 some 25 societies concerned with the environment had formed. Local issues have galvanized support among young people who are disenchanted with state-level environmental politics.[22]

Green movements gained recognition in the Baltic republics of Lithuania, Latvia, and Estonia in 1988. The Lithuanian Green Movement comprises scientists, environmentalists, and nationalists. In October 1988, 30,000 Lithuanians formed a "ring of life" around a Chernobyl style nuclear reactor and then successfully petitioned for a shutdown. They have protested against pollution and promoted healthier food. Environmentalism is seen as a way to restore Lithuanian heritage and to press for independence from Moscow. The Green "Panda" ecology program in Estonia protests air pollution (in Sillamäe workers must wear gas masks) and water pollution (in Tapa one can set fire to drinking water). It promotes natural limits, green education, and green conscience.[23]

EARTH FIRST!

Bonnie and Doc

> parked the car out of sight of the highway, on a turnoff, and walked the half mile back to their objective. The usual precautions. As usual he carried the chain saw, she led the way (she had better night vision). They stumbled through the dark, using no other light than that of the stars, following the right of way fence. . . . They came to the target. It looked the same as before.
>
> MOUNTAIN VIEW RANCHETTE ESTATES
> TOMORROW'S NEW WAY OF LIVING TODAY!
> Horizon Land & Development Corp.
>
> "Beautiful," she said, leaning against the panting Doc.
>
> "Beautiful," he agreed. After resting a moment he put down his McCulloch, knelt, turned on the switch, set the choke, grasped the throttle and gave a good pull on the starter cord. The snappy little motor buzzed into life; the wicked chain danced forward in its groove. He stood up, the machine vibrating in his hands, eager for destruction. He pushed the oiler button, revved the engine and stepped to the nearest upright post of the billboard.

So begins an undercover action of two eco-raiders, immortalized in Edward Abbey's novel, *The Monkeywrench Gang* (1975).[24]

Inspired by Abbey, and founded by disenchanted environmentalist Dave Foreman in the early 1980s, Earth First! advocates strategic ecotage. Its bible is *Ecodefence: A Field Guide to Monkeywrenching* and its medium of communication is *Earth First!: The Radical Environmental Journal*, which proclaims "no compromise in defense of Mother Earth." Although in 1990 they divided into two factions with two separate journals, Earth First!ers are not a organized movement in a formal sense. Rather they are a loose association of "earth warriors" dedicated to saving wilderness through sabotaging the machines that destroy it. They are furious at the failure of the Forest Service and the Bureau of Land Management (BLM) to set aside America's last heritage of wilderness and at bureaucratic environmentalism for lack of aggressive action. Its methods are demonstrations, guerrilla theater, civil disobedience, and monkeywrenching. Many in the movement consider themselves anarchists and all deny that they have been responsible for any injuries to human beings.

To Earth First! the results of the Forest Service's 1977–8 RARE II (Roadless Area Review and Evaluation) survey were scandalous. Out of 80 million acres of national forests with a total area equivalent to the size of New Mexico, only 15 million acres—too high, dry, cold, or steep for logging—were slated for protection. Old-growth forests in the northwest were fingered for logging. A subsequent study DARN (Development Activities in Roadless Non-selected) recommended construction of nine thousand miles of logging roads. The BLM survey identified sixty million acres or an area about the size of Oregon. Of these, only about nine million acres will probably be recommended for wilderness status.

Ecodefense calls on women and men to act individually and in small groups to defend the wild. "Strategic monkeywrenching . . . can be effective in stopping timber cutting, road building, overgrazing, oil and gas exploration, mining, dam building, powerline construction, off-road-vehicle use, trapping, ski area development, and other forms of destruction of the wilderness. . . . But it must be strategic, it must be thoughtful, it must be deliberate in order to succeed." The manual stresses that monkeywrenching is nonviolent and should not be directed at human beings or other living things. "It is aimed at inanimate machines and tools." Monkeywrenching should not be used when other forms of nonviolent confrontation are in progress such as blockades or other forms of direct-action civil-disobedience since it could result in backlash against the protesters or undercut delicate negotiations.[25]

Earth First! direct actions include blockades of logging roads, tree-sits in old-growth forests, demonstrations outside of US Forest Service offices, lumber company sit-ins, and protests over the Smithsonian Institute's proposal to place an observatory in Arizona threatening the habitat of the Mount Graham red squirrel. In more notable nonviolent actions Earth First!ers have padlocked themselves to bulldozers, locked themselves to the cranes of log export ships to support US millworkers, entered their own grazing protest floats in ranchers' livestock parades, and scaled coliseum walls with protest banners.[26]

But Earth First! has a philosophy that goes beyond simple direct actions in defense of the wilderness. For Foreman, it is essential that people maintain their evolutionary ties to the wild. "I am a product of the Pleistocence epoch, the age of large mammals," he holds. "I do not want to live in a world without jaguars and great blue whales and redwoods and rain forests, because this is my geological era, this is my family, this is my context. I only have meaning *in situ*, in the age I live in, the late Pleistocene." Accordingly, the *Earth First!* journal devotes much attention to issues such as old-growth forests, tropical rainforest deforestation, bear and wolf habitat, and endangered species.[27]

But Foreman also embraces deep ecology as a philosophy of nature and a Malthusian view of population. (Earth First! prints bumper stickers that say "Malthus was Right!") Not all Earth First!ers would agree with this position, nor would all deep ecologists. Yet in the Earth First! journal, columnist Miss Ann Thropy writes, "I take it as axiomatic that the only real hope for the continuation of diverse ecosystems on this planet is an enormous decline in human population . . . if the AIDS epidemic didn't exist, radical environmentalists would have to invent one." In 1990, Dave Foreman apologized for this statement calling it an insensitive remark.[28]

Foreman has also been quoted on Ethiopia as stating that "the best thing would be to just let nature seek its own balance, to let the people there just starve." On Latin American immigration he argues, "Letting the USA be an overflow valve for problems in Latin America is not solving a thing. It's just putting more pressure on the resources we have in the USA . . . and it isn't helping the problems in Latin America." Such views were attacked by social ecologists Murray Bookchin and George Bradford for being racist and élitist.[29]

A reconciliation of divisions between social ecologists and Earth

First!ers was initiated in a 1989 debate that included Bookchin and Foreman and a subsequent book by the two protagonists. Bookchin declared solidarity with Earth First!, emphasizing his own love of wilderness. Marxism, he argued, does not go far enough in questioning the domination of nature by humans. Yet a feeling of compassion for other human beings is also necessary if we are to express our feelings of compassion for nature. Foreman, on his part, cited multinational greed as a root cause of social injustice. Seeing the earth as a natural resource to be exploited for profits leads to seeing humans in the same way.[30]

GREENPEACE

Using direct action and confrontation as strategies for change, Greenpeace, now the largest international environmental organization, takes on a variety of issues, from promoting nuclear-free seas, to saving whales and seals, to protesting the waste trade and toxics, and saving Antarctica. Beginning in 1971, it used the Quaker tradition of bearing personal witness to atrocities, such as sailing into Pacific nuclear test areas, expanding the strategy to global witnessing through the mass media.

Greenpeace started its "save the whales" campaign in 1973 when a New Zealand biologist working in Vancouver liberated a killer whale from an aquarium. In 1975 volunteers in a rubber boat confronted Soviet whaling harpooners, capturing the event on film. After that moment of international recognition, it confronted whaling countries through the International Whaling Commission and organized boycotts and grassroots rallies. After ten years of protest only three countries, Japan, Iceland, and Norway continue to harvest whales. Action against these holdouts continues and against the Whaling Commission's relaxation of whaling bans and quotas.

Subsequent Greenpeace campaigns against seal hunting, joined by animal liberation movements, sharply curtailed the international trade in fur, but also drew criticism from seal hunters who pointed out that their original subsistence economies, which captured seals only out of need, were converted to the fur trade by the same western forces now again depriving them of their livelihoods. The Greenpeace campaign for dolphin-safe tuna, caught by lines rather than miles of plastic nondegradable driftnets that trap dolphins and other nontarget species, was supported by San Francisco's Earth Island Institute. In response to such concerns, expressed worldwide, the United Nations issued a ban on driftnet fishing effective in 1993.

In the Pacific, after taking on the issue of nuclear-free seas, international attention was gained when the French sunk a protesting Greenpeace vessel, the Rainbow Warrior, in 1985 in New Zealand. Greenpeace continued to block and tag naval vessels carrying nuclear weapons and reported them through newspapers such as the New York Times. As a result, dozens of ports and nations banned ships carrying such weapons. Greenpeace activists also exposed ships carrying toxic wastes destined for dumps in Third World countries such as Guyana, Guinea, Honduras, the Bahamas, Panama, and Tonga. Seventy-eight countries subsequently banned waste imports. In Antarctica, Greenpeace monitored trash, diesel fuel, and human waste that research stations dumped into the ocean and promoted the idea of a world park instead.[31]

Like Earth First!, Greenpeace has attacked forms of industrial production that threaten the reproduction of life. Thus it fights to save whales, dolphins and seals, and organizes campaigns against nuclear weapons and toxic dumping. Its efforts have thus focused on resolving the contradiction between production and reproduction. Its ethic is fundamentally biocentric—individual life forms, especially those valued by humans and saved through human witnessing, are sacred.

DIRECT ACTION

The "Day After Earth Day," dawns foggy and cool. At 5: 00 AM sleepy activists, still tired from the activities of Earthday 1990, roll over and struggle out of sleeping bags strewn on friends' floors and porches. Gulping down coffee and a little cereal, they carry bicycles, some adorned with hobby-horse heads, down steep San Francisco rowhouse steps, and coast down deserted streets toward the city's hub. They arrive at Pine and Sansome to find an assortment of colorfully dressed bears, birds, and flowers gathering outside the doors of the Pacific Stock Exchange. Posters proclaiming "Liberate the Earth," and "Reforest Corporate Wastelands;" drums, rattles, and megaphones; pots containing dead trees; and chants of "Earth First, Profits Last" announce the sentiments of the demonstrators. On hand, also, are news reporters, with television cameras and camcorders, police officers, and curious onlookers.

As the first suited businessmen and spike-healed businesswomen arrive, the crowd begins to chant, "Earth first, shut it down," and "Don't go to work." The police form a tight semicircle to hold back the nearest demonstrators and escort the stockbrokers toward the entrance. A group of six protesters pushes its way to the door and forms a line across it. They link arms and stand stoically as police arrest them one by one, tie their wrists with plastic handcuffs, and move them toward the waiting police wagons. The crowd roars and presses forward, shouting, "The whole world is watching; the whole world is watching."

Big city police-officers are now accustomed to demonstrations. Direct action groups are now accustomed to the techniques of passive resistance. Both sides have received training in methods of nonviolent confrontation. Evolving from the violent demonstrations of the 1960s and 1970s are codes of behavior and permissible responses to the heated emotions generated by political protests. Demonstrations and marches are scheduled in advance with city authorities, streets are cordoned off, and police are assigned to the area. Demonstrators who agree in advance to be arrested form affinity groups, learn how to block entrances peacefully, how to succumb to or resist arrest in nonviolent ways, and how to support each other and make group decisions at demonstration and holding sites. Supporters who do not wish to be arrested learn how to watch and record the demonstration's progress in order to gather evidence for any court appearances that may result. Volunteer attorneys advise would-be arrestees of their rights and responsibilities and represent them in court.

Inspired by the Gandhian philosophy of nonviolence and the concept of civil disobedience, the direct action movement has developed an array of methods that draw public attention to political issues, but do so in ways designed to minimize bodily harm. Nonviolence preparation preceeding a planned action may involve several hours of training. Such sessions typically cover:

- The history and philosophy of direct action and nonviolence, including role plays on the use of nonviolence and nonviolent responses to violence.
- Role plays and exercises in decision making, conflict resolution, and quick decision making.
- A presentation on the legal ramifications of civil disobedience, and discussion on noncooperation and bail solidarity.
- Exercises and discussion of the role of social oppression and the progressive movement.
- What is an affinity group and what are the roles within the group?
- A sharing of fears and feelings related to nonviolence and nonviolent action.[32]

The direct action movement evolved out of the civil rights sit-ins and antiwar demonstrations of the 1960s and 1970s. Environmental groups began to use the techniques to protest nuclear-power plants in the 1970s. The Clamshell Alliance that demonstrated for many years against New Hampshire's Seabrook nuclear reactor, and the west coast Abalone Alliance that protested the construction and start-up of California's Diablo nuclear power plant further developed the method. Other actions such as the Women's Pentagon Action (1980), the Livermore Action Group (1982 and 1983), and the Rocky Flats, Colorado actions of the mid-1980s used nonviolence to protest nuclear-weapons research and funding. The Pacific Stock Exchange Action in San Francisco was planned in coordination with a similar Wall Street Action by a coalition that included Greens, left Greens, the National Toxics Campaign, and the Environmental Project on Central America in order to make visible the central role played by "banks, stock traders, insurance operators and corporate headquarters" in an economy that "profits from destroying forests, building nuclear weapons, and poisoning our food and water." Earth First!s' Redwood Summer (1990), organized to protest the cutting of the last old-growth redwood forests in the northwestern United States, was similarly committed to nonviolence and held advance training sessions on how to respond to confrontations with the timber industry.[33]

Nonviolent direct action has had both successes and failures in achieving its goals. On the positive side, its methods bring public attention to the issues, since demonstrations often make newspaper and television headlines. The movement pushes the dialogue further to the left, so that organizations such as the Sierra Club and the National Wildlife Federation appear more centrist. It raises people's consciousness about issues so that more moderate initiatives and legislation have a better chance of passing. Some protests have been very successful both in planning, implementation, and subsequent solidarity among the participants and arrestees. On the negative side, because the large crowds that may be attracted to a planned nonviolent action are difficult to control, demonstrations may deteriorate into unfocused, unruly, and even violent occasions where people may be harmed and property damaged. This tends to detract from the overall goal and arouses public resentment.

CONCLUSION

The decade of the 1990s finds an environmental movement that is vigorous, yet diverse and deeply factionalized. Many different groups have sprung up, organized around a multiplicity of causes. Major divisions exist between mainstream politics and movement activism, white majorities and ethnic minorities, conservatives and radicals, wilderness preservationists and humanists. Most mainstream environmental groups, such as the Group of Ten, work within established structures of governance that reproduce the social order, pushing them to repair the problems of production by passing new laws to clean up

the environment and preserve open spaces. Green parties and movements, however, advocate new forms of governance, especially at the
community level, that will reproduce society in ecologically responsible ways.

For much of the toxics and minority environmental movements,
human health and welfare problems are rooted in the malign side-
effects of industrial-capitalist development. These groups try to resolve
the contradictions between production and reproduction so that all
people, especially minorities and the poor, will reproduce their daily
lives in healthy neighborhoods with a reasonable standard of living.
Their homocentric approach seeks a resolution of environmental problems that will benefit the underclasses (women, minorities, wage laborers, and Third World peoples), either through tighter regulation of the
externalities of production or a major restructuring of the economy
itself.

For Earth First!ers, many Greenpeace activists, and deep ecologists, the welfare of wilderness and other species has priority over, or
equal to, the welfare of humans. These groups try to resolve the
contradictions between production and reproduction that prevent animals, plants, and other living things from reproducing themselves
within their own local ecosystems. These biocentric and ecocentric
approaches see humans as only one part of nature, ideally a much
smaller part, than that occupied by present populations. Yet the seemingly separate issues of wilderness and people's health are beginning
to merge, as toxic dumps and radioactive wastes poison the wilderness
and inner cities are made healthier through restoring "the wild" within
their boundaries. Whether the various groups can build coalitions that
offer mutual support for each other's issues, thinking globally, yet
acting locally, is a question yet to be answered.

FURTHER READING

Abbey, Edward. *The Monkeywrench Gang*. New York: Avon, 1975.

———. *Desert Solitaire*. New York: Simon and Schuster, 1968.

Bahro, Rudolf. *Building the Green Movement*. London: New Society Publishers, 1986.

————. *From Red to Green*. London: Basil Blackwell, 1984.

Bullard, Robert D. *Dumping in Dixie: Race, Class, and Environmental Quality*. Boulder, Co.: Westview Press, 1990.

Capra, Fritjof and Charlene Spretnak. *Green Politics: The Global Promise*. New York: E. P. Dutton, 1984.

Chase, Steve, ed. *Defending the Earth: A Dialogue Between Murray Bookchin and Dave Foreman*. Boston, Ma.: South End Press, 1991.

Cohen, Gary and John O'Connor, eds. *Fighting Toxics: A Manual for Protecting Your Family, Community, and Workplace*. Washington, D. C.: Island Press, 1990.

Die Grünen: Programme of the German Green Party. East Haven, Ct.: LongRiver Books, 1983.

Davis, John and Dave Foreman. *The Earth First! Reader: Ten Years of Radical Environmentalism*. Salt Lake City, Ut.: Peregrine Smith Books, 1991.

Epstein, Barbara. *Political Protest and Cultural Revolution: Nonviolent Direct Action in the US*. Berkeley: University of California Press, 1991.

Foreman, David and Bill Haywood, eds. *Ecodefense: A Field Guide to Monkeywrenching*, second ed. Tucson, Az.: Ned Ludd Books, 1987.

Hülsberg, Werner. *The German Greens: A Social and Political Profile*. New York: Verso, 1988.

Hutton, Drew, ed. *Green Politics in Australia*. Sydney: Angus and Robertson, 1987.

Manes, Christopher. *Green Rage*. Boston: Little, Brown, & Co., 1990.

Paehlke, Robert. *Environmentalism and the Future of Progressive Politics*. New Haven, Ct.: Yale University Press, 1989.

Parkin, Sara. *Green Politics: An International Guide*. London: Heretic Books, 1989.

Papadakis, Elim. *The Green Movement in West Germany*. New York: St. Martin's Press, 1984.

Pearce, Fred. *Green Warriors*. London: Bodley Head, 1991.

Pepper, David. *The Roots of Modern Environmentalism*. London: Croom Helm, 1984.

Porritt, Jonathon. *Seeing Green: The Politics of Ecology Explained*. London: Basil Blackwell, 1984.

Scarce, Rik. *Eco-Warriors: Understanding the Radical Environmental Movement*. Chicago: Noble Press, 1990.

Smith, Garry J. *Toxic Cities*. Kensington, NSW: New South Wales University Press, 1990.

Tokar, Brian. *The Green Alternative: Creating an Ecological Future*. San Pedro: R. & E. Miles, 1987.

Weston, Joe, ed. *Red and Green: The New Politics of the Environment*. Wolfeboro, N. H.: Pluto Press, 1986.

8

ECOFEMINISM

In Kenya, women of the Green Belt movement band together to plant millions of trees in arid degraded lands. In India, they join the chipko (tree-hugging) movement to preserve precious fuel resources for their communities. In Sweden, feminists prepare jam from berries sprayed with herbicides and offer a taste to members of parliament: they refuse. In Canada, they take to the streets to obtain signatures opposing uranium processing near their towns. In the United States, housewives organize local support to clean up hazardous waste sites. All these actions are examples of a worldwide movement, increasingly known as "ecofeminism," dedicated to the continuation of life on earth.

Ecofeminism emerged in the 1970s with an increasing consciousness of the connections between women and nature. The term, "ecofeminisme," was coined by French writer Françoise d'Eaubonne in 1974 who called upon women to lead an ecological revolution to save the planet.[1] Such an ecological revolution would entail new gender relations between women and men and between humans and nature.

Developed by Ynestra King at the Institute for Social Ecology in Vermont about 1976, the concept became a movement in 1980 with a major conference on "Women and Life on Earth: Ecofeminism in the '80s," and the ensuing Women's Pentagon Action to protest anti-life nuclear war and weapons development.[2] During the 1980s cultural feminists in the United States injected new life into ecofeminism by arguing that both women and nature could be liberated together.

Liberal, cultural, social, and socialist feminism have all been concerned with improving the human/nature relationship and each has contributed to an ecofeminist perspective in different ways (Table 8.1).[3] Liberal feminism is consistent with the objectives of reform environmentalism to alter human relations with nature from within existing structures of governance through the passage of new laws and regulations. Cultural ecofeminism analyzes environmental problems from within its critique of patriarchy and offers alternatives that could liberate both women and nature.

Social and socialist ecofeminism ground their analyses in capitalist patriarchy. They ask how patriarchal relations of reproduction reveal the domination of women by men, and how capitalist relations of production reveal the domination of nature by men. The domination of women and nature inherent in the market economy's use of both as resources would be totally restructured. Although cultural ecofeminism has delved more deeply into the woman-nature connection, social and socialist ecofeminism have the potential for a more thorough critique of domination and for a liberating social justice.

Ecofeminist actions address the contradiction between production and reproduction. Women attempt to reverse the assaults of production on both biological and social reproduction by making problems visible and proposing solutions (see Figure I.1). When radioactivity from nuclear power-plant accidents, toxic chemicals, and hazardous wastes threatens the biological reproduction of the human species, women experience this contradiction as assaults on their own bodies and on those of their children and act to halt them. Household products, industrial pollutants, plastics, and packaging wastes invade the homes of First World women threatening the reproduction of daily life, while direct access to food, fuel, and clean water for many Third World women is imperiled by cash cropping on traditional homelands and by pesticides used in agribusiness. First World women combat these assaults by altering consumption habits, recycling wastes, and protesting production and disposal methods, while Third World women act to protect traditional ways of life and reverse ecological damage from multinational corporations and the extractive industries. Women challenge the ways in which mainstream society reproduces itself through socialization and politics by envisioning and enacting alternative gender roles, employment options, and political practices.

Many ecofeminists advocate some form of an environmental ethic that deals with the twin oppressions of the domination of women and nature through an ethic of care and nurture that arises out of women's culturally constructed experiences. As philosopher Karen Warren conceptualizes it:

> An ecofeminist ethic is both a critique of male domination of both women and nature and an attempt to frame an ethic free of male-gender bias about women and nature. It not only recognizes the multiple voices of women, located differently by race, class, age, [and] ethnic considerations, it centralizes those voices. Ecofeminism builds on the multiple perspectives of those whose perspectives are typically omitted or undervalued in dominant discourses, for example Chipko women, in developing a global perspective on the role of male domination in the exploitation of women and nature. An ecofeminist perspective is thereby . . . structurally pluralistic, inclusivist, and contextualist, emphasizing through concrete example the crucial role context plays in understanding sexist and naturist practice.[4]

An ecofeminist ethic, she argues, would constrain traditional ethics based on rights, rules, and utilities, with considerations based on care, love, and trust. Yet an ethic of care, as elaborated by some feminists, falls prey to an essentialist critique that women's nature is to nurture.[5]

Table 8.1

Feminism and the Environment

	Nature	Human Nature	Feminist Critique of Environmentalism	Image of a Feminist Environmentalism
Liberal Feminism	Atoms Mind/Body dualism Domination of Nature	Rational Agents Individualism Maximization of self-interest	"Man and his environment" leaves out women	Women in natural Resources and environmental sciences
Marxist Feminism	Transformation of Nature by science and technology for human use. Domination of nature as a means to human freedom Nature is material basis of life: food, clothing, shelter, energy	Creation of human nature through mode of production, praxis Historically specific—not fixed Species nature of humans	Critique of capitalist control of resources and accumulation of goods and profits	Socialist society will use resources for good of all men and women Resources will be controlled by workers Environmental pollution could be minimal since no surpluses would be produced Environmental research by men and women

(continued)

Table 8.1 (continued)

Feminism and the Environment

	Nature	Human Nature	Feminist Critique of Environmentalism	Image of a Feminist Environmentalism
Cultural Feminism	Nature is spiritual and personal Conventional science and technology problematic because of their emphasis on domination	Biology is basic Humans are sexual reproducing bodies Sexed by biology/gendered by society	Unaware of interconnectedness of male domination of nature and women Male environmentalism retains hierarchy Insufficient attention to environmental threats to women's reproduction (chemicals, nuclear war)	Woman/Nature both valorized and celebrated Reproductive freedom Against pornographic depictions of both women and nature Cultural ecofeminism
Socialist Feminism	Nature is material basis of life: food, clothing, shelter, energy Nature is socially and historically constructed Transformation of nature by production and reproduction	Human nature created through biology and praxis (sex, race, class, age) Historically specific and socially constructed	Leaves out nature as active and responsive Leaves out women's role in reproduction and reproduction as a category Systems approach is mechanistic not dialectical	Both nature and human production are active Centrality of biological and social reproduction Dialectic between production and reproduction Multileveled structural analysis Dialectical (not mechanical) systems Socialist ecofeminism

An alternative is a partnership ethic that treats humans (including male partners and female partners) as equals in personal, household, and political relations and humans as equal partners with (rather than controlled-by or dominant-over) nonhuman nature. Just as human partners, regardless of sex, race, or class must give each other space, time, and care, allowing each other to grow and develop individually within supportive nondominating relationships, so humans must give nonhuman nature space, time, and care, allowing it to reproduce, evolve, and respond to human actions. In practice, this would mean not cutting forests and damming rivers that make people and wildlife in flood plains more vulnerable to "natural disasters;" curtailing development in areas subject to volcanos, earthquakes, hurricanes, and tornados to allow room for unpredictable, chaotic, natural surprises; and exercising ethical restraint in introducing new technologies such as pesticides, genetically-engineered organisms, and biological weapons into ecosystems. Constructing nature as a partner allows for the possibility of a personal or intimate (but not necessarily spiritual) relationship with nature and for feelings of compassion for nonhumans as well as for people who are sexually, racially, or culturally different. It avoids gendering nature as a nurturing mother or a goddess and avoids the ecocentric dilemma that humans are only one of many equal parts of an ecological web and therefore morally equal to a bacterium or a mosquito.

LIBERAL ECOFEMINISM

Liberal feminism characterized the history of feminism from its beginnings in the seventeenth century until the 1960s. It is rooted in liberalism, the political theory that accepts the scientific analysis that nature is composed of atoms moved by external forces, a theory of human nature that views humans as individual rational agents who maximize their own self-interest, and capitalism as the optimal economic structure for human progress. It accepts the egocentric ethic that the optimal society results when each individual maximizes her own productive potential. Thus what is good for each individual is good for society as a whole. Historically, liberal feminists have argued that women do not differ from men as rational agents and that exclusion from educational and economic opportunities have prevented them from realizing their own potential for creativity in all spheres of human life.[6]

Twentieth century liberal feminism was inspired by Simone de Beauvoir's *The Second Sex* (1949) and by Betty Friedan's *The Feminine Mystique* (1963). De Beauvoir argued that women and men were biologically different, but that women could transcend their biology, freeing themselves from their destiny as biological reproducers to assume masculine values. Friedan challenged the "I'm just a housewife" mystique resulting from post-World War II production forces that made way for soldiers to reassume jobs in the public sphere, pushing the "reserve army" of women laborers back into the private sphere of the home. The liberal phase of the women's movement that exploded in the 1960s demanded equity for women in the workplace and in education as the means of bringing about a fulfilling life. Simultaneously, Rachel Carson made the question of life on earth a public issue. Her 1962 *Silent Spring* focused attention on the death-producing effects of chemical insecticides accumulating in the soil and tissues of living organisms—deadly elixars that bombarded human and nonhuman beings from the moment of conception until the moment of death.[7]

For liberal ecofeminists (as for liberalism generally), environmental problems result from the overly rapid development of natural resources and the failure to regulate pesticides and other environmental pollutants. The way the social order reproduces itself through governance and laws can be meliorated if social reproduction is made environmentally sound. Better science, conservation, and laws are therefore the proper approaches to resolving resource problems. Given equal educational opportunities to become scientists, natural resource managers, regulators, lawyers, and legislators, women, like men, can contribute to the improvement of the environment, the conservation of natural resources, and the higher quality of human life. Women, therefore, can transcend the social stigma of their biology and join men in the cultural project of environmental conservation.

Within the parameters of mainstream government and environmental organizations, such as the Group of Ten, are a multitude of significant opportunities for women to act to improve their own lives and resolve environmental problems. Additionally, women have established their own environmental groups. Organizations founded by women tend to have high percentages of women on their boards of directors. In California, for example, the Greenbelt Alliance was founded by a woman in 1958, the Save the Bay Association by three women in 1961, and the California Women in Timber in 1975 by a group of women. Yet most of the women in these organizations do not consider themselves feminists and do not consider their cause feminist. Feminism as a radical label, they believe, could stigmatize their long term goals. On the other hand, groups such as Friends of the River, Citizens for a Better Environment, and the local chapter of the Environmental Defense Fund employ many women who consider themselves feminists and men who consider themselves sensitive to feminist concerns, such as equality, childcare, overturning of hierarchies within the organization, and creating networks with other environmental organizations.[8]

CULTURAL ECOFEMINISM

Cultural feminism developed in the late 1960s and 1970s with the second wave of feminism (the first being the women's suffrage movement of the early-twentieth century). Cultural ecofeminism is a response to the perception that women and nature have been mutually associated and devalued in western culture. Sherry Ortner's 1974 article, "Is Female to Male as Nature is to Culture," posed the problem that motivates many ecofeminists. Ortner argued that, cross-culturally and historically women, as opposed to men, have been seen as closer to nature because of their physiology, social roles, and psychology. Physiologically, women bring forth life from their bodies, undergoing the pleasures, pain, and stigmas attached to menstruation, pregnancy, childbirth, and nursing, while men's physiology leaves them freer to travel, hunt, conduct warfare, and engage in public affairs. Socially, childrearing and domestic caretaking have kept married women close to the hearth and out of the workplace. Psychologically, women have been have assigned greater emotional capacities with greater ties to the particular, personal, and present than men who are viewed as more rational and objective with a greater capacity for abstract thinking.[9]

To cultural ecofeminists the way out of this dilemma is to elevate and liberate women and nature through direct political action. Many cultural feminists celebrate an era in prehistory when nature was symbolized by pregnant female figures, trees, butterflies, and snakes and in which women were held in high esteem as bringers forth of life. An emerging patriarchal culture, however, dethroned the mother goddesses and replaced them with male gods to whom the female deities became subservient. The scientific revolution of the seventeenth century further degraded nature by replacing Renaissance organicism and a nurturing earth with the metaphor of a machine to be controlled and repaired from the outside. The ontology and epistemology of mechanism are viewed by cultural feminists as deeply masculinist and exploitative of a nature historically depicted in the female gender. The earth is dominated by male-developed and male-controlled technology, science, and industry.[10]

Often stemming from an anti-science, anti-technology standpoint, cultural ecofeminism celebrates the relationship between women and nature through the revival of ancient rituals centered on goddess worship, the moon, animals, and the female reproductive system. A vision in which nature is held in esteem as mother and goddess is a source of inspiration and empowerment for many ecofeminists. Spirituality is seen as a source of both personal and social change. Goddess worship and rituals centered around the lunar and female menstrual cycles, lectures, concerts, art exhibitions, street and theater productions, and direct political action (web-spinning in anti-nuclear protests) are all examples of the re-visioning of nature and women as powerful forces. Cultural ecofeminist philosophy embraces intuition, an ethic of caring, and web-like human-nature relationships.[11]

For cultural feminists, human nature is grounded in human biology. Humans are biologically sexed and socially gendered. Sex/gender relations give men and women different power bases. Hence the personal is political. The perceived connection between women and biological reproduction turned upside down becomes the source of women's empowerment and ecological activism. Women's biology and Nature are celebrated as sources of female power. This form of ecofeminism has largely focused on the sphere of consciousness in relation to nature—spirituality, goddess worship, witchcraft—and the celebration of women's bodies, often accompanied by social actions such as anti-nuclear or anti-pornography protests.[12]

Much populist ecological activism by women, while perhaps not explicitly ecofeminist, implicitly draws on and is motivated by the connection between women's reproductive biology (nature) and male-designed technology (culture). Many women activists argue that male-designed and produced technologies neglect the effects of nuclear radiation, pesticides, hazardous wastes, and household chemicals on women's reproductive organs and on the ecosystem. They protest against radioactivity from nuclear wastes, power plants, and bombs as a potential cause of birth defects, cancers, and the elimination of life on earth. They expose hazardous waste sites near schools and homes as permeating soil and drinking water and contributing to miscarriages, birth defects, and leukemia. They object to pesticides and herbicides being sprayed on crops and forests as potentially affecting children and child-bearing women living near them. Women frequently spearhead local actions against spraying and power plant siting and organize citizens to demand toxic clean-ups.[13]

In 1978, Lois Gibbs of the Love Canal Homeowner's Association in Niagara Falls, New York, played a critical role in raising women's consciousness about the effects of hazardous waste disposal by Hooker Chemicals and Plastics Corporation in her neighborhood of 1,200 homes. Gibbs, whose son had experienced health problems after attending the local elementary school, launched a neighborhood campaign to close the school after other neighborhood women corroborated her observations. A study conducted by the women themselves found a higher than normal rate of miscarriages, stillbirths, and birth defects. Because the blue collar male population of Love Canal found it difficult to accept the fact that they could not adequately provide for their families, the women became leaders in the movement for redress. Love Canal is a story of how lower-middle-class women who had never been environmental activists became politicized by the life-and-death issues directly affecting their children and their homes and succeeded in obtaining redress from the state of New York. "The women of Love Canal," said Gibbs at the 1980 conference on Women and Life on Earth, "are no longer at home tending their homes and gardens. . . . Women who at one time looked down at people picketing, being arrested, and acting somewhat radical are now doing those very things."[14]

The majority of activists in the grassroots movement against toxics, are women (see Chapter 7). Many became involved when they experienced miscarriages or their children suffered birth defects or contracted leukemia or other forms of cancer. Through networking with neighborhood women, they began to link their problems to nearby hazardous waste sites. From initial Not in My Backyard (NIMBY) concerns, the movement has changed to Not in Anybody's Backyard (NIABY) to Not On Planet Earth (NOPE). Thus Cathy Hinds, whose well water in East Gray, Maine was contaminated by chemicals from a nearby industrial clean-up corporation became "fighting mad" when she lost a child and her daughter began to suffer from dizzy spells. She eventually founded the Maine Citizens' Coalition on Toxics and became active in the National Toxics Campaign. Her motive was to protect her children. Women, she says, "are mothers of the earth," who want to take care of it.[15]

Native American women organized WARN, Women of All Red Nations to protest high radiation levels from uranium mining tailings on their reservations and the high rates of aborted and deformed babies as well as issues such as the loss of reservation lands and the erosion of the family. They recognized their responsibilities as stewards of the land and expressed respect for "our Mother Earth who is a source of our physical nourishment and our spiritual strength."[16]

Cultural ecofeminism, however, has its feminist critics. Susan Prentice argues that ecofeminism, while asserting the fragility and interdependence of all life, "assumes that women and men . . . have an essential human nature that transcends culture and socialization." It implies that what men do to the planet is bad; what women do is good. This special relationship of women to nature and politics makes it difficult to admit that men can also develop an ethic of caring for nature. Second, ecofeminism fails to provide an analysis of capitalism that explains why it dominates nature. "Capitalism is never seriously tackled by ecofeminists as a process with its own particular history, logic, and struggle. Because ecofeminism lacks this analysis, it cannot develop an effective strategy for change." Moreover, it does not deal with the problems of poverty and racism experienced by millions of women around the world.[17] In contrast to cultural ecofeminism, the social and socialist strands of ecofeminism are based on a socioeconomic analysis that treats nature and human nature as socially constructed, rooted in an analysis of race, class, and gender.

SOCIAL ECOFEMINISM

Building on the social ecology of Murray Bookchin, social ecofeminism envisions the restructuring of society as humane decentralized communities. "Social ecofeminism," states Janet Biehl, "accepts the basic tenet of social ecology, that the idea of dominating nature stems from the domination of human by human. Only ending all systems of domination makes possible an ecological society, in which no states or capitalist economies attempt to subjugate nature, in which all aspects of human nature—including sexuality and the passions as well as rationality—are freed." Social ecofeminism distinguishes itself from spiritually oriented cultural ecofeminists who acknowledge a special historical relationship between women and nature and wish to liberate both together. Instead it begins with the materialist, social feminist analysis of early radical feminism that sought to restructure the oppressions imposed on women by marriage, the nuclear family, romantic love, the capitalist state, and patriarchial religion.

Social ecofeminism advocates the liberation of women through overturning economic and social hierarchies that turn all aspects of life into a market society that today even invades the womb. It envisions a society of decentralized communities that would transcend the public-private dichotomy necessary to capitalist production and the bureaucratic state. In them women emerge as free participants in public life and local municipal workplaces.

Social ecofeminism acknowledges differences in male and female reproductive capacities, inasmuch as it is women and not men who menstruate, gestate, give birth, and lactate, but rejects the idea that these entail gender hierarchies and domination. Both women and men are capable of an ecological ethic based on caring. In an accountable face-to-face society, childrearing would be communal; rape and violence against women would disappear. Rejecting all forms of determinism, it advocates women's reproductive, intellectual, sensual, and moral freedom. Biology, society, and the individual interact in all human beings giving them the capacity to choose and construct the kinds of societies in which they wish to live.[18]

But in her 1991 book, *Rethinking Ecofeminist Politics*, Janet Biehl withdrew her support from ecofeminism, and likewise abandoned social ecofeminism, on the grounds that the concept had become so fraught with irrational, mythical, and self-contradictory meanings that it undercut women's hopes for a liberatory, ecologically-sane society. While early radical feminism had sought equality in all aspects of public and private life, based on a total restructuring of society, the cultural feminism that lies at the root of much of ecofeminism seemed to her to reject rationality by embracing goddess worship, to biologize and essentialize the caretaking and nurturing traits assigned by patriarchy to women, and to reject scientific and cultural advances just because they were advocated by men.[19] Social ecofeminism, however, is an area that will receive alternative definition in the future as theorists such as Ynestra King, Ariel Salleh, Val Plumwood, and others sharpen its critique of patriarchal society, hierarchy, and domination. Women of color will bring still another set of critiques and concerns to the ongoing dialogue.

SOCIALIST ECOFEMINISM

Socialist ecofeminism is not yet a movement, but rather a feminist transformation of socialist ecology that makes the category of reproduction, rather than production, central to the concept of a just, sustainable world. Like Marxist feminism, it assumes that nonhuman nature is the material basis of all of life and that food, clothing, shelter, and energy are essential to the maintenance of human life. Nature and human nature are socially and historically contructed over time and transformed through human praxis. Nature is an active subject, not a passive object to be dominated, and humans must develop sustainable relations with it. It goes beyond cultural ecofeminism in offering a critique of capitalist patriarchy that focuses on the dialectical relationships between production and reproduction, and between production and ecology.

A socialist ecofeminist perspective offers a standpoint from which to analyze social and ecological transformations, and to suggest social actions that will lead to the sustainability of life and a just society. It asks:

1. What is at stake for women and for nature when production in traditional societies is disrupted by colonial and capitalist development?
2. What is at stake for women and for nature when traditional methods and norms of biological reproduction are disrupted by interventionist technologies (such as chemical methods of birth control, sterilization, amniocentesis, rented wombs, and baby markets) and by chemical and nuclear pollutants in soils, waters, and air (pesticides, herbicides, toxic chemicals, and nuclear radiation)?
3. What would an ecofeminist social transformation look like?
4. What forms might socialist societies take that would be healthy for all women and men and for nature?

In his 1884 *Origin of the Family, Private Property, and the State*, Friedrich Engels wrote that "the determining factor in history is, in the last resort, the production and reproduction of immediate life. . . . On the one hand, the production of the means of subsistence . . . on the other the production of human beings themselves." In producing and reproducing life, humans interact with nonhuman nature, sustaining or disrupting local and global ecologies. When we ignore the consequences of our interactions with nature, Engels warned, our conquests "take . . . revenge on us." "In nature nothing takes place in isolation." Elaborating on Engels' fundamental insights, women's roles in production, reproduction, and ecology can become the starting point for a socialist ecofeminist analysis.[20]

SOCIALIST ECOFEMINISM AND PRODUCTION

As producers and reproducers of life, women in tribal and traditional cultures over the centuries have had highly significant interactions with the environment. As gatherers of food, fuel, and medicinal herbs; fabricators of clothing; planters, weeders, and harvesters of horticultural crops; tenders of poultry; preparers and preservers of food; and bearers and caretakers of young children, women's intimate knowledge of nature has helped to sustain life in every global human habitat.

In colonial and capitalist societies, however, women's direct interactions with nature have been circumscribed. Their traditional roles as producers of food and clothing, as gardeners and poultry tenders, as healers and midwives, were largely appropriated by men. As agriculture became specialized and mechanized, men took over farm production, while migrant and slave women and men supplied the stoop labor needed for field work. Middle-class women's roles shifted from production to the reproduction of daily life in the home, focusing on increased domesticity and the bearing and socialization of young children. Under capitalism, as sociologist Abby Peterson points out, men bear the responsibility for and dominate the production of exchange commodities, while women bear the responsibility for reproducing the workforce and social relations. "Women's responsibility for reproduction includes both the biological reproduction of the species (intergenerational reproduction) and the intragenerational reproduction of the work force through unpaid labor in the home. Here too is included the reproduction of social relations—socialization." Under industrial capitalism, reproduction is subordinate to production.[22]

Because capitalism is premised on economic growth and competition in which nature and waste are both externalities in profit maximization, its logic precludes sustainability. The logic of socialism on the other hand is based on the fulfillment of people's needs, not people's greed. Because growth is not necessary to the economy, socialism has the potential for sustainable relations with nature. Although state socialism has been based on growth-oriented industrialization and has resulted in the pollution of external nature, new forms of socialist ecology could bring human production and reproduction into balance with nature's production and reproduction. Nature's economy and human economy could enter into a partnership.

The transition to a sustainable global environment and an equitable human economy that fulfills people's needs would be based on two dialectical relationships—that between production and ecology and that between production and reproduction. In existing theories of capitalist development, reproduction and ecology are both subordinate to production. The transition to socialist ecology would reverse the priorities of capitalism, making production subordinate to reproduction and ecology.

SOCIALIST ECOFEMINISM AND REPRODUCTION

Socialist ecofeminism focuses on the reproduction of life itself. In nature, life is transmitted through the biological reproduction of species in the local ecosystem. Lack of proper food, water, soil chemicals, atmospheric gases, adverse weather, disease, and competition by other species can disrupt the survival of offspring to reproductive age. For humans, reproduction is both biological and social. First, enough children must survive to reproductive age to reproduce the community over time; too many put pressure on the particular mode of production, affecting the local ecology. Second, by interacting with external nature, adults must produce enough food, clothing, shelter, and fuel on a daily basis to maintain their own subsistence and sustain the quality of their ecological homes. Both the intergenerational biological reproduction of humans and other species and the intragenerational reproduction of daily life are essential to continuing life over time. Sustainability is the maintenance of an ecological-productive-reproductive balance between humans and nature—the perpetuation of the quality of all life.[22]

Biological reproduction affects local ecology, not directly, but as mediated by production. Many communities of tribal and traditional peoples developed rituals and practices that maintained their populations in a balance with local resources. Others allowed their populations to grow in response to the need for labor or migrated into new lands and colonized them. When the mode of production changes from an agrarian to an industrial base, and then to a sustainable production base, the number of children that families' need declines. How development occurs in the future will help families decide how many children to have. A potential demographic transition to smaller population sizes is tied to ecologically sustainable development.

Ecofeminist political scientist Irene Diamond raises concern over the implications of "population control" for Third World women. "The 'advances' in family planning techniques from Depra-Provera to a range of implanted birth control devices, banned in western nations as unsafe, reduce Third World women to mindless objects and continue the imperialist model which exploits native cultures 'for their own good.'"[23] Second, with the availability of prenatal sex identification techniques, feminists fear the worldwide "death of the female sex" as families that place a premium on male labor opt to abort as many as nine out of every ten female fetuses. Third, feminists argue that women's bodies are being turned into production machines to test contraceptives, for *in vitro* fertilization experiments, to produce babies for organ transplants, and to produce black market babies for sale in the northern hemisphere.

Reproductive freedom means freedom of choice—freedom to have or not to have children in a society that both needs them and provides for their needs. The same social and economic conditions that provide security for women also promote the demographic transition to lower populations. The Gabriella Women's Coalition of the Philippines calls for equal access to employment and equal pay for women, daycare for children, healthcare, and social security. It wants protection for women's reproductive capacities, access to safe contraception, and the elimination of banned drugs and contraceptives. It advocates equal, nondiscriminatory access to education, including instruction concerning consumer rights and hazardous chemicals. Such a program would help to bring about a sustainable society in which population is in balance with the fulfillment of daily needs and the use of local resources,

a society that offers women and men of all races, ages, and abilities equal opportunities to have meaningful lives.

A socialist ecofeminist movement in the developed world can work in solidarity with women's movements to save the environment in the underdeveloped world. It can support scientifically-based ecological actions that also promote social justice. Like cultural ecofeminism, socialist ecofeminism protests chemical assaults on women's reproductive health, but puts them in the broader context of the relations between reproduction and production. It can thus support point of production actions such as the Chipko and Greenbelt movements in the Third World (see below), protests by Native American women over cancer-causing radioactive uranium mining on reservations, and protests by working class women over toxic dumps in urban neighborhoods.[24]

WOMEN IN THE THIRD WORLD

Many of the problems facing Third World women today are the historical result of colonial relations between the First and Third Worlds. From the seventeenth century onward, European colonization of lands in Africa, India, the Americas, and the Pacific initiated a colonial ecological revolution in which an ecological complex of European animals, plants, pathogens, and people disrupted native peoples' modes of subsistence, as Europeans extracted resources for trade on the international market and settled in the new lands. From the late eighteenth century onward, a capitalist ecological revolution in the northern hemisphere accelerated the extraction of cash crops and resources in the southern hemisphere, pushing Third World peoples onto marginal lands and filling the pockets of Third World élites. In the twentieth century, northern industrial technologies and policies have been exported to the south in the form of development projects. Green Revolution agriculture (seeds, fertilizers, pesticides, dams, irrigation equipment, and tractors), plantation forestry (fast-growing, non-indigenous species, herbicides, chip harvesters, and mills), capitalist ranching (land conversion, imported grasses, fertilizers, and factory

farms) and reproductive technologies (potentially harmful contraceptive drugs, sterilization, and bottle feeding) have further disrupted native ecologies and peoples.

Third World women have born the brunt of environmental crises resulting from colonial marginalization and ecologically unsustainable development projects. As subsistence farmers, urban workers, or middle-class professionals, their ability to provide basic subsistence and healthy living-conditions is threatened. Yet Third World women have not remained powerless in face of these threats. They have organized movements, institutes, and businesses to transform maldevelopment into sustainable development. They are often at the forefront of change to protect their own lives, those of their children, and the life of the planet. While some might consider themselves feminists, and a few even embrace ecofeminism, most are mainly concerned with maintaining conditions for survival.

In India, nineteenth century British colonialism in combination with twentieth century development programs has created environmental problems that affect women's subsistence, especially in forested areas. Subsistence production, oriented toward the reproduction of daily life, is undercut by expanding market production, oriented toward profit-maximization (see Figure I.1). To physicist and environmentalist, Vandana Shiva, the subsistence and market economies are incommensurable:

> There are in India, today, two paradigms of forestry—one life-enhancing, the other life-destroying. The life-enhancing paradigm emerges from the forest and the feminine principle; the life-destroying one from the factory and the market. . . . Since the maximising of profits is consequent upon the destruction of conditions of renewability, the two paradigms are cognitively and ecologically incommensurable. The first paradigm has emerged from India's ancient forest culture, in all its diversity, and has been renewed in contemporary times by the women of Garhwal through Chipko.[25]

India's Chipko, or tree-hugging, movement attempts to maintain sustainability. It has its historical roots in ancient Indian cultures that worshipped tree goddesses, sacred trees as images of the cosmos, and

sacred forests and groves. The earliest woman-led tree-embracing movements are three-hundred years old. In the 1970s women revived these chipko actions in order to save their forests for fuelwood and their valleys from erosion in the face of cash cropping for the market. The basis of the movement lay in a traditional ecological use of forests for food (as fruits, roots, tubers, seeds, leaves, petals and sepals), fuel, fodder, fertilizer, water, and medicine. Cash cropping by contrast severed forest products from water, agriculture, and animal husbandry. Out of a women's organizational base and with support by local males, protests to save the trees took place over a wide area from 1972 through 1978, including actions to embrace trees, marches, picketing, singing, and direct confrontations with lumberers and police.[26]

The Chipko movement's feminine forestry-paradigm is based on assumptions similar to those of the emerging science of agroforestry, now being taught in western universities. Agroforestry is one of several new sciences based on maintaining ecologically viable relations between humans and nature. As opposed to modern agriculture and forestry, which separate tree crops from food crops, agroforestry views trees as an integral part of agricultural ecology. Complementary relationships exist between the protective and productive aspects of trees and the use of space, soil, water, and light in conjunction with crops and animals. Agroforestry is especially significant for small farm families, such as many in the Third World, and makes efficient use of both human labor and natural resources.[27]

In Africa, numerous environmental problems have resulted from colonial disruption of traditional patterns of pastoral herding as governments imposed boundaries that cut off access to migratory routes and traditional resources. The ensuing agricultural development created large areas of desertified land, which had negative impacts on women's economy. The farmers, mostly women, suffered from poor yields on eroded soils. They had to trek long distances to obtain wood for cooking and heating. Their cooking and drinking waters were polluted. Developers with professional training, who did not understand the meaning of "development without destruction," cut down trees that interfered with highways and electrical and telephone lines, even if they were the only trees on a subsistence farmer's land.

Kenyan women's access to fuelwood and water for subsistence was the primary motivation underlying the women's Greenbelt Movement. According to founder Wangari Maathai, the movement's objective is to promote "environmental rehabilitation and conservation and . . . sustainable development." It attempts to reverse humanly-produced desertification by planting trees for conservation of soil and water.[28]

The National Council of Women of Kenya began planting trees in 1977 on World Environment Day. Working with the Ministry of the Environment and Natural Resources, they continued to plant trees throughout the country and established community woodlands on public lands. They planted seedlings and sold them, generating income. The movement promoted traditional agroforestry techniques that had been abandoned in favor of "modern" farming methods that relied on green revolution fertilizers, pesticides, new seed varieties, and irrigation systems that were costly and non-sustainable. During the past ten years, the movement has planted over seven million trees, created hundreds of jobs, reintroduced indigenous tree species, educated people in the need for environmental care, and promoted the independence and a more positive image of women.[29]

"The whole world is heading toward an environmental crisis," says Zimbabwe's Sithembiso Nyoni. "Women have been systematically excluded from the benefits of planned development. . . . The adverse effects of Africa's current so-called economic crisis and external debt . . . fall disproportionately on women and make their problems ever more acute." Twenty years ago there was still good water, wood, grass, and game even on semi-arid communal lands and women did not have to walk long distances to obtain subsistence resources. But the introduction of Green Revolution seeds and fertilizers required different soils and more water than found on the common lands. The poor, primarily women, have born the brunt of development that has proceeded independently of environmental consequences.[30]

According to Zimbabwe's Kathini Maloba, active in both the

Greenbelt Movement and the Pan-African Women's Trade Union, many farm women suffer loss from poor crops on marginal soils, lack of firewood, polluted water, poor sanitation, and housing shortages. Women have suffered miscarriages from the use of chemical fertilizers and pesticides. In 1983, 99 percent of all farms had no protection from pesticides. Only 1 percent of employers heeded pesticide warnings and used detection kits to test pesticide levels in foods and water.

Development programs that emphasize people's needs within local environmental constraints would include: water conservation through erosion control, protection of natural springs, and the use of earthen dams and water tanks; in agriculture, the reintroduction of traditional seeds and planting of indigenous trees; in herding, the use of local grasses, seeds, and leaves for feed and driving cattle into one place for fattening before market; in homes, the use of household grey water to irrigate trees and more efficient ovens that burn less fuelwood.

Latin American women likewise point to numerous environmental impacts on their lives. Both Nicaragua and Chile are countries in which socialist governments have been opposed by the United States through the use of economic boycotts and the funding of opposition leaders who supported conservative capitalist interests. Maria Luisa Robleto of the Environmental Movement of Nicaragua asserts that women are fighting to reverse past environmental damage. In Nicaragua, before the Sandinista revolution of 1979, many women worked on private haciendas that used large amounts of pesticides, especially DDT. Since the revolution the postion of women changed as part of the effort to build a society based on sustainable development. In part because of male engagement in ongoing defense of the country and in part because of the efforts of the Nicaraguan women's movement, women moved into agricultural work that was formerly masculine. Women were trained in tractor driving, coffee plantation management, and animal husbandry.

According to Robleto, women agricultural workers in Nicaragua have twenty times the level of DDT in their breast milk as non-agricultural workers. They want equal pay and an end to toxic poisoning from insecticides. If breast feeding is promoted as an alternative to expensive formula feeding, there must be a program to control toxics in breast milk. In a country where 51 percent of the energy comes from firewood, 39 percent of which is used for cooking, there must be a forestry and conservation program oriented to women's needs. A grassroots movement is the spark for ecological conservation.

Chile's Isabelle Letelier of the Third World Women's Project (widow of the Chilean ambassador to the United States who was assassinated in 1976 by Pinochet agents after the 1973 overthrow of the socialist Allende government), speaks of the power of *compesina* women who created life and controlled medicine and religion. The global society, she says, is out of control. The round planet must be saved. Women must take charge, since men are not going to solve the problems. They must construct a society for both women and men. The rights of the land, the rights of nature, and women's rights are all part of human rights. Santiago is now one of the most polluted cities in the world. There are children who receive no protein and who resort to eating plastic. There is a television in every home, but no eggs or meat. There are colored sugars, but no bread. In 1983, says Letelier, women broke the silence and began speaking out for the environment. Without the help of telephones, they filled a stadium with 11,000 women. They established networks as tools; they learned to question everything, to be suspicious of everything. They learned to see. "Women give life," says Letelier. "We have the capacity to give life and light. We can take our brooms and sweep the earth. Like witches, we can clean up the atmosphere with our brooms. We can seal up the hole in the ozone layer. The environment is life and women must struggle for life with our feet on the ground and our eyes toward the heavens. We must do the impossible."

Gizelda Castro, of Friends of the Earth, Brazil, echoes the ecofeminist cry that women should reverse the damage done to the earth. "Men," she says, "have separated themselves from the ecosystem." Five hundred years of global pillage in the name of development and civilization have brought us to a situation of international violence against the land and its people. The genetic heritage of the south is constantly going to the north. Women have had no voice, but ecofeminism is a new and radical language. Women must provide the moral energy and determination for both the First and Third Worlds. They are the future and hope in the struggle over life.[31]

In Malaysia, which received independence in 1957 as the British empire underwent decolonization, many environmental problems have resulted from a series of five-year development plans which ignored both the environment and conservation, especially the impact of development on women. "The rapid expansion of the cash crop economy which is hailed as a 'development success story' has plunged thousands of women into a poisonous trap," argues Chee Yoke Ling, lecturer in law at the University of Malaysia and secretary general of the country's chapter of Friends of the Earth. As land control shifted to large multinational rice, rubber, and palm oil plantations, women's usufructory rights to cultivate the land were lost to a male-dominated cash-exporting economy. They became dependent and marginalized, moving into low paying industrial and agricultural jobs. Women workers constitute 80 percent of those who spray chemical pesticides and herbicides such as paraquat on rubber and palm plantations. They pour the liquid, carry the open containers, and spray the chemicals without protective clothing, even when pregnant or nursing. The workers are usually unaware of the effects of the chemicals and often cannot read the warning labels on the packaging. Protests resulted in loss of jobs or transfer to even less desirable forms of labor. In 1985, Friends of the Earth Malaysia began to pressure the Ministry of Health to ban paraquat. They called on plantation owners and government agencies to stop using the chemical for the sake of human right to life as well as the life of waters and soils.[32]

Third World women are thus playing an essential role in conservation. They are making the impacts of colonialism and industrial capitalism on the environment and on their own lives visible. They are working to maintain their own life-support systems through forest and water conservation, to rebuild soil fertility, and to preserve ecological diversity. In so doing, they are assuming leadership roles in their own communities. Although they have not yet received adequate recognition from their governments and conservation organizations for their contributions, they are slowly achieving the goals of ecofeminism—the liberation of women and nature.

WOMEN IN THE SECOND WORLD

Second World development has been informed by Marxist theory that the goal of production is the fulfillment of human needs. Yet state socialism as the method for achieving equitable distribution of goods and services has created enormous problems of pollution and depletion resulting from a series of five-year plans for rapid industrial growth (see Chapter 1). As Second World countries incorporate market economic goals, environmental problems will become increasingly complex. Can the evolving, changing Second World produce and distribute enough food and goods for its own people and also reverse environmental deterioration? The movements toward democratization in the 1990s reveal an openness to new ideas and cooperation in resolving economic and environmental problems, but many problems in implementing solutions remain.

While Second World women have shared educational and economic opportunities along with men, like First World women they have also borne the double burden of housework added to their employment outside the home. Like First World women, they have experienced the effects of industrial and toxic pollutants on their own bodies and seen the impacts on their children and husbands. Although women in the Second World have not achieved the environmental vision of Marxist feminists (see Table 8.1), they have used scientific and technological research and education to find ways of mitigating these problems and have participated in incipient green movements.

Second World women have assumed leadership roles in environmental affairs. In Poland, Dr. Maria Guminska, a professor of biochemistry at Krakow Medical University helped to found the 4000 member Polish Ecology Club and served as one of its vice-presidents. She prepared a critical report on the air pollution of Poland's largest aluminum smelter and was active in the effort to reduce toxic pollutants from a Krakow pharmaceutical plant.

In the former Soviet Union, Dr Eugenia V. Afanasieva, of the Moscow Polytechnical Institute, was Deputy Director of the Environmental Education Center for Environmental Investigation. The Center developed a filtration system to help clean up industrial water pollution. Dr Afanasieva works with young people to promote better environmental education. "All mankind now stands at the beginning of a new era," she states. "People must make the choice to live or to perish. Nobody can predict the future. We must save our civilization. We must change our ways of thinking. We must think ecologically." Women, she argues, play a major role in expanding environmental awareness: "It seems to me that women are more active in environmental programs than men. We give birth to our children, we teach them to take their first steps. We are excited about their future."[33]

In 1989 the First International Conference on Women, Peace, and the Environment was held in the former Soviet Union. The women called for greater participation by women as environmentalists and scientists to help decide the fate of the planet. They said:

> Each of us should do everything possible to promote actions for survival on local, national, and international levels. . . . We must work to end food irradiation, to ban all known chemicals destroying the ozone layer, to reduce transport emissions, to recycle all reusable waste, to plant arboreta and botanical gardens, to create seed banks, etc. These are among the most urgent beginnings for a strategy of survival.[34]

Olga Uzhnurtsevaa of the Committee of Soviet Women pleads for environmental improvement in the face of her country's accelerating industrial production. A national ecological program subsidized by the government is needed to reverse ecological damage. Children are being born with birth defects; air and water quality have deteriorated. Throughout the Commonwealth, she says, women's councils support environmental thinking. Many of the journalists and activists concerned over environmental problems in the Lake Baikal watershed and the Baltic Sea are women. Women are especially concerned with the need to protect nature from the arms race. This problem involves all of humanity, especially the effects on the Third World. Quoted Uzhnurtsevaa,

> Nature said to women:
> Be amused if you can,
> Be wise if possible,
> But by all means, be prudent.[35]

CONCLUSION

Although the ultimate goals of liberal, cultural, social, and socialist feminists may differ as to whether capitalism, women's culture, or socialism should be the ultimate objective of political action, shorter-term objectives overlap. Weaving together the many strands of the ecofeminist movement is the concept of reproduction construed in its broadest sense to include the continued biological and social reproduction of human life and the continuance of life on earth. In this sense there is perhaps more unity than diversity in women's common goal of restoring the natural environment and quality of life for people and other living and nonliving inhabitants of the planet.

FURTHER READING

Biehl, Janet. *Rethinking Ecofeminist Politics*. Boston: South End Press, 1991.

Carson, Rachel. *Silent Spring*. Boston: Houghton Mifflin, 1962.

Dankelman, Irene and Joan Davidson. *Women and Environment in the Third World*. London: Earthscan Publications, 1988.

Diamond, Irene and Gloria Orenstein, eds. *Reweaving the World: The Emergence of Ecofeminism*. San Francisco: Sierra Club Books, 1990.

Eisenstein, Zillah, ed. *Capitalist Patriarchy and the Case for a Socialist Feminism*. New York: Monthly Review Press, 1979.

Gilligan, Carol. *In a Different Voice: Psychological Theory and Women's Development*. Cambridge, Mass.: Harvard University Press, 1982.

Jaggar, Alison. *Feminist Politics and Human Nature*. Totowa, New Jersey: Rowman and Allanheld, 1983.

King, Ynestra. "Feminism and the Revolt of Nature," Heresies, 4, no. 1 (1981):12–16.

———. "Toward an Ecological Feminism and a Feminist Ecology," in Joan Rothschild, ed., *Machina Ex Dea*. New York: Pergamon Press, 1983, pp. 118–129.

Kuhn, Annette and Ann Marie Wolpe, eds. *Feminism and Materialism: Women and Modes of Production*. London: Routledge and Kegan Paul, 1978.

Merchant, Carolyn. *The Death of Nature: Women, Ecology, and the Scientific Revolution*. San Francisco: Harper and Row, 1980.

Mies, Maria. *Patriarchy and Accumulation on a World Scale*. London: Zed Books, 1986.

Momsen, Janet Henshall. *Women and Development in the Third World*. New York: Routledge, 1991.

Noddings, Nel. *Caring, a Feminine Approach to Ethics and Moral Education*. Berkeley: University of California Press, 1984.

O'Brien, Mary. *The Politics of Reproduction*. Boston: Routledge and Kegan Paul, 1981.

Ortner, Sherry. "Is Female to Male as Nature is to Culture?" in *Woman, Culture, and Society*, ed. Michelle Rosaldo and Louise Lamphere. Stanford: Stanford University Press, 1974, pp. 67–87.

Plant, Judith. *Healing the Wounds: The Promise of Ecofeminism*. Philadelphia, Pa.: New Society Publishers, 1989.

Seager, Joni and Ann Olson. *Women in the World: An International Atlas*. New York: Simon and Schuster, 1988.

Shiva, Vandana. *Staying Alive: Women, Ecology and Development*. London: Zed Books, 1988.

Waren, Karen, ed. "Ecological Feminism," special issue of *Hypatia*, 6, no. 1 (Spring 1991).

9

SUSTAINABLE DEVELOPMENT

Don Jose Jesus Mendoza leans on his spade and surveys his plot of land. "People thought I was crazy, " he muses, "when they saw me mixing weeds with manure, water, and dirt. But when they saw I doubled my harvest last year, then they wanted to know how I did it." Mendoza is a sixty year old Nicaraguan farmer, carpenter, and poet who is part of an active *Compesino* to *Compesino* (farmers teaching farmers) movement in Central America. An inventive, enthusiastic teacher, he tells his fellow farmers, "*Companeros*, this course begins with two words and ends with two words: 'Organic Matter' and 'Organic Matter.'"[1]

Sustainable agriculture is part of a larger program of sustainable development (SD) oriented to converting ecologically destructive production into environmentally sound production. Unlike green politics and ecofeminism, which act to resolve the contradiction between production and reproduction, the sustainability movement attempts to resolve the contradiction between production and ecology by making production ecologically sustainable. Like the green and ecofeminist movements, however, the SD movement is diverse, containing within it a spectrum of political approaches and ethical orientations (see Table 3.1). Most organic farmers, ecological restorationists, and bioregionalists see humans as one part of an ecological web and implicitly employ an ecocentric land ethic. Mainstream sustainable development (to some people a contradiction in terms), as exemplified by the 1987 United Nations (Bruntland) report, *Our Common Future*, is homocentric and utilitarian in its approach.[2] The SD movement is informed by both deep ecological and socialist ecological theory. It has not yet incorporated much ecofeminist theory although it is consistent with many of the goals of ecofeminists.

SUSTAINABLE AGRICULTURE

Sustainable agriculture, as practiced by Nicaragua's Don Jose Mendoza, is an ecologically based form of farm management. Soil is a living thing. Feeding the soil, rather than feeding the plant alone, builds long-lasting fertility. Using biological processes maintains and improves the soil, whereas pesticides and herbicides degrade it. Synthetic fertilizers may serve the fertilizer industry rather than the soil. Excessive use of chemical inputs contaminates ground water. Instead, intensive management by the farmer working in harmony with nature optimizes yields. Compost, crop rotations, diversification, polycultures, cover crops, and careful selection of varieties lead to better tasting, nutritious products. Crops are selected for local markets, rather than for resistance to shipping damage, and for local climate and soil conditions, rather than for standardized green revolution seeds and technologies.

Before using sustainable agriculture, Nicaraguan peasants had employed the age-old method of slash and burn. They cleared land with fire and planted crops for two or three years in the nutrient rich soil. But as more people needed more land there was not enough to let the land lie fallow for the ten to twenty years needed to recover its fertility. They farmed the same soil without fertilizing or protecting it and dreamed of owning large amounts of land. "Now I know how to work the land," says Mendoza, "I'm just fine with my seven manzanas (approximately 12 acres)."

Mendoza's course is part of a Central American program in which farmers teach each other new sustainable methods of agriculture. Local agencies send their best *compesino* promoters and agricultural technicians to *Compesino* Development Centers to share knowledge and practices with each other. They then help to teach the techniques to their comrades back home. Because the *compesino* to *compesino* program is low-cost, and labor-intensive, it works well in agrarian communities where farmers have access to small plots of land.[3]

Sustainable agriculture is posited in opposition to industrialized agriculture, which is based on optimizing purchased inputs to produce outputs at the least cost. The "evolution from labor intensive to energy and capital intensive farming," says Miguel Altieri of the University of California at Berkeley, "was not influenced by rational decisions based on ecological considerations, but mainly by the low cost of energy inputs." In contrast, the ecological approach is based on principles that conserve the renewable resource base and reduce the need for external technological inputs. Scientists argue that sustainability can be achieved through ecological methods that incorporate the wisdom of traditional peoples.

According to Gordon Douglass of Pomona College in southern California, the principles of sustainable agriculture include:

1. The optimization of farm output over a much longer time period than is usual in industrial farming activities.
2. The promotion and maintenance of diversified agroecosystems whose living components perform complementary functions.
3. The building up of soil fertility with organic matter and the protection of nutrients from leaching.

4. The promotion of continuous cover and the extensive use of legume-based rotations, cover crops, and green manures.
5. The limiting of inported fertilizer applications and pesticide uses.[4]

In achieving sustainability, a systems approach is needed. A particular cover crop may add nitrogen and keep down dust and insects, but encourage nematodes in the soil. By retaining water, it may lower the temperature of an orchard or field and add to frost risk. Thus each change in the transition from high chemical inputs to natural methods needs to be evaluated in the context of the whole agroecosystem, rather than through a reductionist single component approach.[5]

Sustainable agriculture can be further extended to integrate the human community with the agroecosystem. "This holistic approach to farming communities," Douglass points out, "draws attention to interactions not only within [and] among farming families and other human member[s] of rural communities, but also between nonhuman components such as crops with crops, crops with animals, soil conditions and fertility with insects, and disease in crops and livestock." Sustainable agriculture is thus based on an ecocentric ethic of management in which the land is considered as a whole, its human components being only one element. Policy decisions must be based on considerations of what is best for the soil, vegetation, and animals (including humans) on the farm as well as outside sources of water, air, and energy. As a result, humans and the land can be sustained together.[6]

Permaculture, as envisioned by Australians David Holmgren and Bill Mollison, carries sustainabilty a step further. This method of agriculture imitates ecosystem evolution toward climax states through perennial plant and animal crop interactions. In contrast to monocultural agriculture, permaculture uses several stories of trees, shrubs, vines, and perennial ground crops to absorb more light and nutrients, increasing the total yield. Plants and animals coexist in separate niches that reduce competition and promote symbiosis among species. Complexity not only helps to ward off catastrophes, but increases the variety of foods produced. External energy and physical labor decrease as perennials mature, so that energy needs are provided from within the system. Permaculture is highly adaptable and is applicable to a spectrum of habitats from tropical rainforests in Malaysia to arid deserts in Africa.[7]

In Salina Kansas, Wes Jackson devotes his Land Institute to research and experimentation on perennial grains. Horrified by the loss of soil in the most productive lands of the United States, he sees in perennials the hope of saving soil, energy, and time in the fields. The goal is to find and breed perennial grasses that can produce high yields each year, and be planted in polycultures that reduce insects, pathogens, and weeds, and renew soil fertility, especially nitrogen and carbon. Researchers have planted 4000 wild relatives of annual grains in order to isolate hardy high yielding varieties that can be developed through further cross breeding. While the research is still experimental, a few promising grasses and legumes have emerged that could lead to sustainable ecosystem-based agriculture.[8]

Sustainable agriculture is a growing worldwide movement. It is supported by international research and funding efforts, through university research and cooperative extension programs, and by local farmers. Yet sustainable agriculture must also be integrated with farmworker rights that promote social justice and protection from exposure to pesticides and herbicides.

BIOLOGICAL CONTROL

The biological control of insects is a related example of sustainable management. Using ecological guidelines, natural insect enemies are introduced into the ecosystem to control population levels of pests. Uncultivated land surrounding fields harbors birds and pest enemies. Flowers along roadsides and fences are especially attractive to beneficial insects. Diversity in crops and surroundings and arrangements of beneficial plants mimic natural conditions. Crops become less visible to insect enemies and act as a barrier to the spread of pests.

The technique was pioneered in California in 1888. The cottony-cushion scale, introduced from Australia, was destroying citrus groves in southern California. Albert Koebele traveled to Australia and brought back the vedalia, a lady beetle that fed on the scale. One thousand beetles soon cleared acres of orange groves, saving the industry. This ecological strategy was vindicated in the 1940s when DDT killed so many of the vedalia that a resurgence of the scale occurred.[9]

The assumptions that underlie biological control and its related strategy, integrated pest management (IPM), are ecologically grounded. They contrast with chemical control, which assumes that humans are above nature and can legitimately use pesticides to obliterate insects for human benefit. "Biological control, together with plant resistance," writes IPM founder Carl Huffaker of the University of California, "are the core around which pest control in crops and forests should be built." Ecology provides the model for insect control. According to biologist Ray Smith, "we must understand Nature's methods of regulating populations and maximize their application."[10]

Biological control and IPM assume that humans are only one part of an interrelated ecological complex and that insects and humans must coexist. Insect populations will not be totally obliterated, but their numbers can be controlled so that humans may harvest crops. Reservoirs of insect pests, however, will continue to exist. This ecological interdependence implies that all organic and inorganic parts of the ecosystem have intrinsic value. Biological control is based therefore on an ecocentric ethic. This contrasts with the homocentric ethic of chemical-control techniques to manage insects. Chemical control assumes that humans are the most important parts of a complex social and natural world and can manipulate that world for the good of society.[11]

RESTORATION ECOLOGY

A parking lot in California teams with blue-jeaned, tee-shirted volunteers. Shovels, buckets, trash containers, and day packs are scattered around on the ground while people hear the instructions of an ecologist. They have come together for a weekend outing to help restore a parkland newly purchased by the state. They enrich the soil with redwood chips and remove debris and the remnants of an old lumber operation. Guided by ecological principles, they plant young trees, ferns, huckleberries, and ground cover. Their plantings reintroduce the native species that will promote the ecological conditions under which insect, mammal, and bird communities can regenerate themselves. A new whole is created, helping to recreate the major elements of the presettlement ecosystem.[12]

Restoration is the process of restoring human-disturbed ecosystems to earlier pristine forms. It is the active reconstruction of pristine ecosystems (such as prairies, grasslands, rivers, and lakes). By studying and mimicking natural patterns, the wisdom inherent in evolution can be reestablished.

Using ecological guidelines, species are planted according to their original distributions in close proximity to each other. Over time a process occurs in which synergistic relationships are reestablished among soils, plants, insect pollinators, and animals to recreate the prairie ecosystem. Like a doctor healing a patient or a helmsperson steering a boat, restoration is a process of synthesis in which humans put non-human nature back together again. It contrasts with the mechanistic model in which nature is like a clock that can be taken apart through analysis and repaired through external intervention. People can thus live symbiotically within the whole.[13]

But restoration need not apply just to parks and natural areas. Forests, deserts, wetlands, and even cities can be rehabilitated to be ecologically compatible with human uses. Biological principles are used to select fruits and nuts that can be harvested from rainforests allowing economic sustainability. Wetlands can be reconstructed by engineers and replanted by biologists. Indigenous trees and succulents can restore human-created deserts to human-sustainable biosystems. Even cities can become ecocities by uncovering underground creeks and rehabilitating shorelines, marshes, and springs. Urban gardening in backyards and on rooftops, greenbelt areas for wildlife, and forest/meadow and water/land border zones can be created.

BIOREGIONALISM

Bioregionalists are local caretakers. Dedicated to the concept of living-in-place, they espouse "watershed consciousness." They urge that everyone know the source of their local water—where it comes from and were it goes, the hills and valleys into which it flows, and the creeks that lead it to rivers. How many people, rural and urban alike, know the type of soil on which their home is built, the names of even a few native plants and birds, and the mating seasons of local wild animals? How many know the way of life of the tribal peoples that preceded them, how they used the land, and what they gave back to it? Yet passing the bioregional quiz (Table 9.1) with a respectable score, is only the beginning of bioregional consciousness.

"Bioregions," writes Peter Berg (to whom the term is credited), "are geographic areas having common characteristics of soil, watersheds, climate, and native plants and animals that exist within the whole planetary biosphere as unique and intrinsic contributive parts." But beyond the geographical terrain is a terrain of consciousness— ideas that have developed over time about how to live in a given place. Bioregionalism differs from a regional politics of place in its emphasis on natural systems. It includes all the interdependent forms and processes of life, along with humans and human consciousness. "Bioregionalism," observes Jim Dodge, "is simply biological realism; in natural systems we find the physical truth of our being, the real obvious stuff like the need for oxygen as well as the more subtle need for moonlight, and perhaps other truths beyond those."[14]

The roots of bioregionalism go back to the early ecological concept of the biome system of classification, developed by Frederic Clements and Victor Shelford in the 1930s. Biomes were natural habitats such as grasslands, deserts, rainforests, and coniferous forests shaped by climate. Particular soils, vegetation, and animal life developed in each climatic region in accordance with rainfall, temperature, and weather patterns. In the 1970s, Raymond Dasmann, one of the founders of the bioregional movement, helped to redraw the global map in terms of its biotic provinces for the purposes of conservation of plants and animals. He then went on to distinguish between ecosystem people, who for millennia lived within and were dependent on the local ecosystem for survival, and modern biosphere people, who exploit the entire globe for trade in products, breaking down watershed and ecosystem constraints.[15]

Table 9.1

Where You At? A Bioregional Quiz

What follows is a self-scoring test on basic environmental perception of place. Scoring is done on the honor system, so if you fudge, cheat, or elude, you also get an idea of where your're at. The quiz is culture-bound, favoring those people who live in the country over city dwellers, and scores can be adjusted accordingly. Most of the questions, however, are of such a basic nature that undue allowances are not necessary.

1. Trace the water you drink from precipitation to tap.
2. How many days till the moon is full? (Slack of two days allowed.)
3. What soil series are you standing on?
4. What was the total rainfall in your area last year (July–June)? (Slack: 1″ for every 20″.)
5. When was the last time a fire burned your area?
6. What were the primary subsistence techniques of the culture that lived in your area before you?
7. Name five native edible plants in your region and their season(s) of availability.
8. From what direction do winter storms generally come in your region?
9. Where does your garbage go?
10. How long is the growing season where you live?
11. On what day of the year are the shadows the shortest where you live?
12. When do the deer rut in your region, and when are the young born?
13. Name five grasses in your area. Are any of them native?
14. Name five resident and five migratory birds in your area.
15. What is the land-use history of where you live?
16. What primary geological event/process influenced the land form where you live? (Bonus special: what's the evidence?)
17. What species have become extinct in your area?
18. What are the major plant associations in your region?
19. From where you're reading this, point north.
20. What spring wildflower is consistently among the first to bloom where you live?

Scoring
  0–3 You have your head in a hole.
  4–7 It's hard to be in two places at once when you're not anywhere at all.
  8–12 A fairly firm grasp of the obvious.
  13–16 You're paying attention.
  17–19 You know where you're at.
  20 You not only know where you're at, you know where it's at.

By Leonard Charles, Jim Dodge, Lynn Milliman, and Victoria Stockley. *Co-Evolution Quarterly*, Winter 1981, now known as the *Whole Earth Review*, reprinted by permission.

Table 9.2

The Bioregional Paradigm and the Industrial Scientific Paradigm

	Bioregional Paradigm	Industrial Scientific Paradigm
Scale	Region Community	State Nation/World
Economy	Conservation Stability Self-sufficiency Cooperation	Exploitation Change/Progress World Economy Competition
Polity	Decentralization Complementarity Diversity	Centralization Hierarchy Uniformity
Society	Symbiosis Evolution Division	Polarization Growth/Violence Monoculture

Source: Kirkpatrick Sale, *Dwellers in the Land: The Bioregional Vision*, Philadelphia, Pa.: New Society Publishers, 1991, p. 50. Reprinted by permission.

Bioregionalism advocates a new ecological politics of place. It starts with "bundles" of materials describing a bioregion and its history—maps, native species lists, ecological studies, histories, stories, poems, and celebrations of the inhabitants' ways of life. From its roots in the Planet Drum Foundation in San Francisco in the 1970s, bioregionalism has grown to some 70–100 local North American groups whose addresses are their own bioregions. Annual gatherings in different watersheds around the country bring people together to develop and share strategies for change.

Knowing the land, learning the lore, developing the potential, and liberating the self are the tasks of the would-be bioregionalist as seen by Kirkpatrick Sale. Using the human and natural resources of a place entails ecological constraints. The local community is the best body to keep development within the guidelines of human–nature reciprocity. Through this participatory process, one draws closer to other members of the human community. The values inherent in the industrial-scientific paradigm and the bioregional paradigm stand in marked contrast (Table 9.2). The industrial model is neither timely nor sane, but outdated and irrevelant. The bioregional project is neither romantic, utopian, nor nostalgic, but realistic. The problems of moving from the former to the latter are both ecological and political. The changes will be gradual, rather than sudden and revolutionary. They will depend on education, organization, and activism. But if carefully planned, introduced, and implemented, they will be steady, continuous, and truly transformative.[16]

Living within the resources of the local bioregion is one platform in the new politics of place. Using local water and energy sources, bioregional communities attempt to grow their own food and distribute it locally. Dovetailing with the restoration movement, they reconstruct rivers and creeks to support fish runs and clean up and restock lakes. Green city projects attempt to establish reciprocal relations between downstream consumer-dependent cities and upstream rural-producing farmlands and forests. City meetings bring together garbage collectors, industrial scrap companies, and recycling centers with park planners, neighborhood associations, and poets.[17]

The truly bioregional city, Sale argues, must be truly ecological. "The city would have to be as fully rooted in the earth, as close to the natural processes, as the farm and the village." This means growing food in community gardens and farmbelts, producing energy from solar collectors and wind generators, recycling solid and organic wastes, planting trees for producing oxygen and absorbing noise and dust, using mass transit systems, bicycles, and feet, and constructing buildings and homes out of local materials. It means returning organic composts and wastes to farms for reuse and bringing farm products back to the city for sale.[18]

But bioregionalism has its skeptics. Focusing on the neighborhood may preclude seeing the global context; emphasizing the native may obliterate the significance of the introduced, including humans. Ignoring the aquaducts that bring in water and the sewers that carry it away, the air systems that link one city's wastes to another's illnesses, and the imported plants and animals from all parts of the globe oversimplifies the real-world life-equation. "The truth is," lampoons critic Walter Truett Anderson, "that any concept of a 'natural' ecosystem is only a snapshot of how things were at some arbitarily-chosen point of time.

And if you do begin to pay attention to the artificial and new and exotic aspects of your environment, then you have opened yourself up to the contemplation of a world that is much, much more complex than the bioregionalists would have us believe. This is the real world, the world we live in now and are going to have to understand and deal with wisely." Despite such criticism, bioregionalism offers a program of change toward a sustainable way of life. As such it shares many of the goals of indigenous peoples.[19]

INDIGENOUS PEOPLES AND SUSTAINABILITY

Native peoples around the world are drawing on the concept of sustainable management as they attempt to preserve their ways of life. The indigenous Maori of New Zealand are both defending their land from environmental assaults and moving toward a sustainable form of land use. Their movement draws meaning from their traditional origin story of the earth mother and the forest:

> In the beginning there was the Nothingness. Then the Sky Father was joined by the Earth Mother and they had many children who in the darkness between their parents, craved for light. . . . Tane-mahuta god of the forests [said], 'It is better to tear them apart so that we may live with our Mother the Earth and be a stranger to our Father the Sky' . . . . So with sorrow and with tears he tore them apart, and with tears and sorrow heard their lamentations that he, their son, was destroying their great love. But Tane-mahuta was growth, and none could stay him until he had thrust the sky his Father far above; and there he stayed. And to this day, when you see the mists rise from the valleys, you will know that the Earth Mother still sighs for her lost love, while in the still morning his tears fall as the gentle dew.[20]

Today the Maori traditions are represented by the Maori Secretariat, in the Ministry for the Environment. The name of their movement, *Maruwhenua*, comes from the traditional Maori land ethic of human responsibility for shielding the land. The name reflects the saying, "People perish, but the land endures."

The Maori approach to sustainability begins with the following observations:

> Maori people have been denied the authority to influence decisions affecting them regarding resource use. Environmental degradation has seen Maori men and women often the worst affected. . . . The western industrialized approach has brought some good things to the world, but now a number of people are saying that the benefits have not been evaluated in terms of what's been lost . . . our heritage, language, and relationship between us and the land. . . . To address these global issues by starting at home, we need a strong contribution from Maori . . . one that reflects the cultural and spiritual values of the [land].[21]

In the rainforests of Sarawak, Malaysia, Penan gatherer–hunters are engaged in a desperate struggle to retain their way of life. Over five million hectares of rainforest have been licensed for logging for export. Rainforest products are used for paper bags, toilet paper, shipping crates, and furniture by northern hemisphere consumers, who are unaware that each throwaway carton and each roll of toilet paper represents a violation of human and environmental rights in the southern hemisphere. Intensive logging in the Penan lands began during the 1980s, when logging activities shifted from peninsular Malaysia to the states of Sarawak and Sabah, formerly part of the British colony of Borneo, which joined the Malaysian federation in 1963. Timber concessions were given to politicians and ex-civil servants who became wealthy beneficiaries of the political economy of timber.

Occupying the upper tributaries of the Big Baram River (which flows into the South China Sea near the Sarawak town of Miri), the 9000 Penan, traditionally nomadic gatherer–hunters are now mainly semi–settled shifting cultivators. Penan society is gentle and egalitarian. Women and men are equal participants in production, using the forest for sago palm, fruits, bearded pig, deer, monkey, fish, and rattan, all maintained for sustainable use. But logging has disrupted patterns of stewardship, destroying sago, fruit, and rattan patches, food sources for pig and monkey, and fishing rivers. Eroded hillsides, muddied rivers, compacted soil, and barren clearings form huge scars on the land (see cover illustration).

After several years of failure to retain their lands through negotiation, the Penan gathered together in 1987 to stop the bulldozers and loggers from further advance. Men, breast-feeding mothers, and children walked across the mountains for days to join in creating roadblocks across logging routes. Their vigils, supported internationally by rainforest action groups, disrupted logging for months until broken up by police. By the end of 1990, most of the Penan had lost the battle to save their homelands and all but a few families (about 300 people) had abandoned their nomadic way of life to become semi-settled agriculturalists. As the market economy encroaches on the Penan's subsistence economy, the Penan people are being transformed from independent communities to wage laborers in the logging industry and objects of curiosity for the tourist industry in Sarawak's Mulu National Park.

The Penan are supported by rainforest action groups who are trying to pass legislation in First World countries such as Japan, Australia, and the United States to ban imports of rainforest timbers from nonsustainable, primary, and unlogged tropical forests. Second, rainforest groups have urged consumers not to purchase furniture, construction materials, chopsticks, and other products made of rainforest timbers such as meranti, or Pacific maple. Third, they have asked for support for rainforest people so that they may establish sustainable, environmentally and economically sound industries and new farming methods based on permaculture and sustainable agriculture. Fourth, a United Nations Biosphere Reserve has been proposed for the area. Clearly, development in regions such as Sarawak will continue. Whether that development is environmentally sustainable and respectful of the rights and wishes of people such as the Penan is an issue that requires new forms of cooperation and negotiation among indigenous peoples, industries, governments, and environmentalists.[22]

Similarly, Amazonian forest peoples are trying to show that development can respect the way of life of traditional peoples without destroying nature. Until his murder in December 1988 by rainforest clearcutters, Chico Mendes had worked to organize Amazonian rubber tappers. Taught to read newspapers and listen to the radio in the depths of the forest, Mendes led a struggle for social justice. In 1976 he and other rubber tappers marched into the forest and joined hands to stop crews from clearing the rubber trees. Women and children joined in the stand-off. "On at least four occasions we were arrested and forced to lie on the ground with them beating on us." he said. "They threw our bodies, covered in blood, into a truck. We got to the police station and we were a hundred people. They didn't have enough room to keep us there so in the end they had to let us go free." By 1989 the tappers estimated that they had saved some three million acres of rubber trees through the stand-off movement.

Realizing the long-term limitations of the stand-off movement, the tappers began to press for a new legal status for the lands as "extractive reserves." The lands would be given use rights and collective long term leases by the state. Having obtained legal status, people could then organize schools, health clinics, and rubber processing stations. The movement for sustainable management was joined by native Indians who had historically been enemies of the tappers.

The Indigenous People's Union, founded in 1980, began lobbying for Amazonian Indian rights. They demanded that they participate in any development decisions on their lands. They put forward ways to sustain their lands and ways of life. They protested the construction of two dams that would destroy the livelihoods of the Indians along with those of fishers and forest products extractors.

Together, the Alliance of Forest Peoples and the National Rubber Tappers' Council called for a role in designating areas of rubber and brazilnut trees for development without destruction. They argued that the tropical forests could be used as extractive reserves for commercial products without cutting them down or degrading them. The reserves would be under the direct control of the users. Beyond this they called for resettlement of their native lands, an end to rent payments, and the rehabilitation of degraded lands.[23]

In 1990, Brazil established four reserves amounting to 8,351 square miles, roughly the size of the state of Massachusetts, with long term extractive rights for rubber tappers and Indians. As areas in which logging is prohibited, they are being used for extracting nuts, roots, oils, fruits, and pigments. Most of the products are marketed to companies in North America and Europe which use the nuts and oils in rainforest products. The products include new consumer items such as assai-flavored sherbet, cupuacu yogurt, babacu oil, patchouli-root soap, copaiba shampoo, priprioca perfume, and Amazonian latex condoms. The areas are estimated to be twice as profitable per hectare as cattle ranching and the soils will not degrade from clear-cutting.[24]

In Hawaii, the ancient volcanic goddess, Madame Pele is being defended by native Hawaiians who hope to preserve the United States' last tropical rainforest. Pele is active nature, both giver and taker of land. Her violent erruptions expand the island; her lava flows take back the soil from settlers. Hawaiian priests and priestesses gave her fruit and flowers. Hawaiians still offer her the first fruits of the 'ohelo berries that grow on high lava fields. In the Polynesian origin story the earth mother and sky father were united. From them the gods were born— the male gods of the ocean, of humans, and of agriculture and healing, and the female goddesses of fertility, of women's works and of humans. People and all living things were related. Mana, the energy of the world, descended from the godly ancestry to human families. Pele, who seduced her older sister's husband, was driven out of her homeland and crossed the sea to Hawaii, guided by her older brother in the form of a great shark. She went down the Hawaiian chain until she finally made her home in Kilauea volcano on the big island of Hawaii.[25]

Today the Pele Defense Fund is attempting to preserve the Big Island environment from geothermal development. The Hawaiian Electric Company hopes to harness steam from Pele's Kilauea volcano to produce electricity for Hawaii's future development. Native Hawaiians who formed the Pele Defense Fund argue that the drilling would violate their goddess Pele's sacred sanctuary. "Drilling is . . . a sacrilege, no different than trashing a Christian cathedral." The roads, power plants, and transmission line swaths would destroy the delicate ecology of one of the last large tropical rainforests in the United States. Geothermal advocates point to the greater harm of oil burning power plants and the need to free the Hawaiian islands of dependence on foreign oil.[26]

As an alternative to using the land for geothermal development, the Rainforest Action Network proposes energy conservation. They argue that this is a cheaper method with far less impact on the land. If the state became an active participant in the efficiency revolution, it would be five times cheaper than the proposed geothermal plant and would save twice as much electricity. Improved lighting using compact fluorescent lamps, solar water heaters, more efficient refrigerators, and water-saving shower heads could save 68 percent of the energy used by private residences. Imported oil is refined primarily for use in jet airplanes (42 percent) and automobiles (20 percent), rather than for electricity (34 percent). Establishing strong building codes for future development will prevent further energy leaks in an already leaky energy tub. Energy efficiency combined with solar and wind energy would be adequate to meet the state's future energy needs.[27]

Can the knowledge of indigenous peoples be combined with sustainable economic development in Third World countries? Can the visions and programs of local communities be incorporated into national and international development plans? These questions were posed by the United Nations.

GLOBAL SUSTAINABLE DEVELOPMENT

In 1983, the United Nations formed the World Commission on Environment and Development and charged it with preparing "a global agenda for change." Headed by Norwegian Prime Minister, Gro Harlem Bruntland, who had reached her position through years of political struggle as an environmental minister, it produced a major report, *Our Common Future,* in 1987. The Commission, comprising world leaders from some twenty-two countries, sought wide input from organizations and individuals around the world. Its report discussed issues of population, food, species preservation, energy, industry, urbanization, and peace. It called for a new form of economic development that would sustain the resource base.

"Humanity has the ability to make development sustainable," declared the Bruntland Report, "to ensure that it meets the needs of the present without compromising the ability of future generations to meet their needs. . . . Sustainable development requires meeting the basic needs of all and extending to all the opportunity to fulfill their aspirations for a better life." To do this, population and growth must harmonize with the potentials and constraints of the ecosystem. Sustainable development will be the result of difficult choices, policies, institutions, and political will.[28]

The Commission argued that beneficial economic growth will depend on two conditions: (1) the sustainability of the ecosystems involved in exchange, and (2) equity and an end to dominance in the basis for exchange. Sustainable growth in developing countries has been prevented by the debt burdens of Third World countries, whose trade profits must service debt rather than development, and by international projects that have brought short term profits while causing environmental destruction. World Bank and International Development Association projects should support long-term social goals and environmentally sound projects. Development assistance should be directed toward small projects with grassroots cooperation such as: sustainable-regenerative—rather than chemically-dependent—agriculture, reforestation, fuelwood development, watershed protection, soil conservation, agroforestry, rehabilitation of irrigation projects, small-scale agriculture, and low-cost sanitation measures.

Along with its recommendations in such areas as population, food, and energy, the Commission made a number of specific recommendations on how to achieve sustainable economic growth. International commodity trade agreements could be improved in several crucial respects:

1. Larger sums for compensatory financing to even out economic shocks would help to mitigate overproduction of commodities where production is close to the limits of environmental sustainability.
2. More assistance should be given to diversification from single-crop production and for promoting resource regeneration and conservation.
3. More of the environmental and resource costs associated with production should be reflected in the prices of goods produced in developing countries.
4. When transnational corporations introduce new technologies, plants, processes, or joint ventures into developing countries, they should adhere to codes that deal explicitly with the objectives of environmentally sustainable development.

5. Mission-oriented cooperative research ventures in developing countries should be focused on technologies that apply to dryland agriculture, tropical forestry, pollution control, and low-cost housing.[29]

While the Bruntland report has received much praise for its comprehensive examination of global environmental problems, its emphasis on a growth-oriented industrial model of development has been criticized by some developing countries. For example, a group of non-governmental organizations (NGOs) in Paraguay concluded that it emphasized the scientific knowledge of the west over indigenous forms of knowledge and did not appreciate the fact that research funded by the industrialized nations and multinational corporations would tend to favor their own interests rather than those of developing peoples. It proposed instead that development proposals should be judged according to three criteria: (1) improvement of people's lives in both quantitative and qualitative terms, (2) protection of the ecological and cultural heritage of a region, (3) helping the growth of citizens organizations. Rather than aiming for a higher level of economic well-being for all, the world's rich should settle for smaller incomes so that the material conditions of the poor could improve.

A group of Latin American representatives who met in Mexico in October 1987 to evaluate the report agreed that the Bruntland Commission preferred the cultural and economic perspectives of the industrialized nations. They recommended that new models of industrial development be considered and that the United Nations Environmental Program give priority to regional programs.[30]

The Canadian Green Web newsletter went even further. In criticizing the growth-oriented perspective of the report, it argued that a true sustainable development would call for "a massive global transfer of wealth and the cancellation of third-world debts. . . . Environmental protection also means an internal redistribution of productive wealth. . . . We live in a global ecological commons, and the solutions to the rapidly developing disaster we all face have to be global in nature." It criticized the report's human-centered perspective that advocated conscious choices to save or eliminate particular species. The report's "resourcist" worldview implied that "other species do not have intrinsic value in their own right, but are considered 'resources' for human use."[31]

Lester Brown, founder of the Worldwatch Institute in Washington, envisions the sustainable society of 2030. If sustainability means "the capacity to satisfy current needs without jeopardizing the prospects of future generations," this entails: "protecting the ozone layer, stabilizing climate, conserving soils, stabilizing forests and population." By 2030, either sustainability will have been achieved or society will be in a process of continuing disintegration. Existing technologies and energy efficiency are Brown's keys to stabilizing environmental deterioration. Energy will be based on a solar-powered economy in which neither fossil fuels nor nuclear power play a major role—solar panels will be on every rooftop for water heating, and electricity will be supplied by solar power, hydropower, geothermal power, and wind energy. All products will be extraordinarily energy efficient; mass transit and bicycles will be the major transportation methods; agroforestry and small-scale farms will conserve soil; waste reduction and recycling will apply to *all* materials and will replace garbage disposal and land fills. New solar-based and recycling jobs will supersede fossil-fuel based jobs. Global population will have stabilized by 2030 at about eight billion. Values will be based less on material goods and more on fulfillment of human potentials. A transition to sustainability would require a major mobilization of policies, funding, and human energy, but the current global awareness makes that achievement possible.[32]

Both development-oriented and technological approaches to sustainability have been criticized. Economist Lori Ann Thrupp, sees the sustainability movement as split into two main camps (Table 9.3). The dominant group includes northern hemisphere scientists and protectionists who are primarily white, male, upper-middle class, educated professionals and who are employed by well-endowed mainstream environmental organizations, development agencies, banks, private consulting-firms, and universities. These groups are strongly oriented toward wilderness and species preservation, technological solutions, and population control. They tend to devalue social problems such as poverty, lack of housing, garbage and toxic waste disposal in poor areas and Third World countries, and worker health issues. The second

Table 9.3

The Environmental Sustainability Movement

Issues/Aspects	Ecological/Scientific Environmentalism	Anti-establishment Environmentalism
a. View of nature and ecology	Strict preservation; eco-centrism Nature/wildlife protectionism	Natural resources as basis of production
b. Theory and explanation of problems	Functionalist/systems analysis, science prevails, technocentric Causes seen as bad values, greed; poor education; overpopulation	Structural analysis Socio-political roots Capital exploits nature (sometimes Marxist)
c. Ethics/ideology on human-nature	Biological determinism, lifeboat ethics	Equality, social justice, non-exploitation
d. Political view	Liberal to conservativism	Leftist or radical
e. View of people and population	Elitist; superiority of educated scientists; Belief in Mathusianism Poor are ignorant & biased	Emphasis on ill-distribution & exploitation of poor Anti-Malthusianism views Poor are innocent victims
f. Main concerns & focii	Habitat, Wilderness, Energy, ecosystems, species extinction overpopulation, carrying capacity	Sustenance, human environ., slums, toxic waste, worker health
g. Patterns of participation	Narrow; ecoscientists rule; & state policy-makers decide	Grass-roots mobilization Empowerment of poor
h. View of energy problems	Insufficiency & poor tech. & strict limits	Capitalist relations create disparities
i. View of economy & growth	Anti-growth, steady state or controlled growth	Growth not main problem; capitalism is blamed
j. Views about nuclear power	Polluting, inefficient & requires safety laws	Exploitative, tied to capitalism, military-indus. establishment
k. Strategies to overcome problems	Consciousness-raising Preservation/Protection EPA, DOE, technocracy Approp. Technology Education, training Force birth control Romanticize nature Scientists provide fixes	Structural tranfer; overthrow capital that exploits nature Socialist paths; labor movements; political action; social equity Feminist values; justice in resource distribution

Source: Lori Ann Thrupp, "The Political Economy of the Sustainable Development Crusade: From Elite Protectionism to Social Justice," presented at the 1990 Annual Meeting of the Association of American Geographers, Toronto, April, 1990, printed by permission of the author.

group comprises First and Third World grassroots groups, indigenous peoples' movements, anti-establishment greens, urban minority groups, and a few academics, all of whom stress social justice in land, health, education, and quality of life.

To both sides sustainable development (SD) has taken on the characteristics of a crusade, with SD replacing and encompassing 1970s nomenclatures such as appropriate technology, ecodevelopment, integrative rural development, and soft energy paths. Thrupp criticizes the mainstream SD movement as proposing well-intentioned, but over-simplified panaceas such as Third World park preserves, debt-for-nature swaps, population controls, and resettlement of peoples from fragile to less fragile ecozones, rather than addressing northern hemisphere causes such as exploitative investments, over-consumption, trickle-down fund and technology transfers, and lack of law enforcement. Instead she proposes progressive strategies that hear and empower poor majorities, grassroots groups, and indigenous peoples and support diversity in agroecosystems, economic products, and institutions. These should not be idealized or romanticized, but be directly supported through funds and material resources. At the top, centralized institutions can halt the fetishism of economic growth based on conventional economic indicators and GNP, introduce qualitative dimensions into development models, stop subsidizing resource-exploiting sectors, and enforce long-term conservation investments.[33]

CONCLUSION

The sustainability movement encompasses mainstream and grassroots environmental organizations, scientists and political activists, and First and Third World concerns and peoples. It has the imprimateur of the United Nations at the top of the global hierarchy and the *compesino* to *compesino* movement at the bottom. It has characteristics at one extreme of maintaining the *status quo* and at the other of radical, structural, social and environmental change. The sustainability movement has the potential for transforming the conditions of production to make them ecologically viable. Is sustainability a viable option for meaningful transformation? Or is it another passing fad? The answer lies somewhere along the spectrum of possibilites and will depend in large part on the extent to which policies, labor, and funding are redirected toward progressive economic and political priorities.

FURTHER READING

Altieri, Miguel. *Agroecology*. Boulder, Co.: Westview Press, 1987.

Apple, J. L., and R. F. Smith, eds. *Integrated Pest Management*. New York: Plenum Press, 1976.

Berg, Peter. *Figures of Regulation: Guides for Re-Balancing Society with the Biosphere*. San Francisco: Planet Drum Foundation, n.d.

——, ed. *Reinhabiting a Separate Country: A Bioregional Anthology of Northern California*. San Francisco: Planet Drum Foundation, 1978.

Berger, John J., ed. *Environmental Restoration: Science and Strategies for Restoring the Earth*. Washington, D. C.: Island Press, 1990.

Berger, John J. *Restoring the Earth*. New York: Knopf, 1985.

Berry, Wendell. *The Unsettling of America: Culture and Agriculture*. New York: Avon, 1977.

Burger, Julian. *The Gaia Atlas of First Peoples: A Future for the Indigenous World*. New York: Doubleday Anchor, 1990.

Chavasse, C. G. R., and J. H. Johns. *The Forest World of New Zealand, Realm of Tane-mahuta*. Wellington, N. Z.: A. H. & A. W. Reed, 1975.

Debach, Paul. *Biological Control By Natural Enemies*. London: Cambridge University Press, 1974.

Douglass, Gordon K., ed. *Agricultural Sustainability in a Changing World Order*. Boulder, Co.: Westview Press, 1984.

Jackson, Wes. *New Roots for Agriculture*, 2nd ed. Lincoln: University of Nebraska Press, 1985.

Kane, Herb Kawainui. *Pele, Goddess of Hawai'i's Volcanoes*. Captain Cook, Hi.: The Kawainui Press, 1987.

Mendes, Chico. *Fight for the Forest*. London: Latin American Bureau, 1989.

Mollison, Bill, and David Holmgren. *Permaculture One: A Perennial Agriculture for Human Settlements*. Maryborough, Australia: Dominion Press-Hedges and Bell, 1978.

Mollison, Bill. *Permaculture Two: Practical Design for Town and Country in Permanent Agriculture*. Maryborough, Australia: Dominion Press-Hedges and Bell, 1979.

Perkins, John. *Insects, Experts, and the Insecticide Crisis: The Quest for New Pest Management Strategies.* New York: Plenum Press, 1982.

Redclift, Michael. *Development and the Environmental Crisis: Red or Green Alternatives?* New York: Methuen, 1984.

———. *Sustainable Development: Exploring the Contradictions.* New York: Methuen, 1987.

Sale, Kirkpatrick. *Dwellers in the Land: The Bioregional Vision.* Philadelphia, Pa.: New Society Publishers, 1991.

Strange, Marty. *Family Farming: A New Economic Vision.* Lincoln: University of Nebraska Press, 1988.

Soulé, Michael, ed. *Conservation Biology: The Science of Scarcity and Diversity.* Sunderland, Ma.: Sinauer, 1986.

Van den Bosch, Robert. *The Pesticide Conspiracy.* New York: Doubleday Anchor, 1980.

Willers, Bill, ed. *Learning to Listen to the Land.* Washington, D. C.: Island Press, 1991.

World Commission on Environment and Development. *Our Common Future.* Oxford: Oxford University Press, 1987.

Young, John. *Sustaining the Earth.* Kensington, NSW: University of New South Wales Press, 1991.

CONCLUSION:
THE RADICAL
ECOLOGY MOVEMENT

What has the radical ecology movement accomplished? A broad range of answers to this question is possible. Radical ecology has not brought about a worldwide socialist order. Nor is such a scenario likely in the immediate future. Its achievements are far more modest. As a theoretical critique of the mainstream environmental movement, it exposes social and scientific assumptions underlying environmentalists' analyses. As a movement, it raises public consciousness concerning the dangers to human health and to nonhuman nature of maintaining the status quo. In so doing, it pushes mainstream society toward greater equality and social justice. It offers an alternative vision of the world in which race, class, sex, and age barriers have been eliminated and basic human needs have been fulfilled.

What analyses and concrete results have radical theorists and activists contributed to the environmental movement?

CONTRIBUTIONS OF RADICAL THEORISTS

- Reality is a totality of internally related parts. The relationships are fundamental and continually shape the totality as contradictions and conflicts arise and are resolved.
- Social reality has structural (ecological and economic) and superstructural (law, politics, science, and religious) features. Continual change is generated out of the contradictions and interactions among the parts and levels.
- Science is not a process of discovering ultimate truths of nature, but a social construction that changes over time. The assumptions accepted by its practitioners are value-laden and reflect their places in both history and society, as well as the research priorities and funding sources of those in power.
- Ecology is likewise a socially constructed science whose basic assumptions and conclusions change in accordance with social priorities and socially accepted metaphors.
- What counts as a natural resource is historically contingent and is dependent on a particular cultural and economic system in a given place and time.
- Surplus and scarcity are produced by economic interactions with non-human nature. Scarcity is both real in that some resources are non-rewewable over human lifespans and created in that economic producers control the technologies of extraction and the distribution of commodities.
- Human reproduction is not determined by indiscriminate sexual passions, but is governed by cultural norms and practices.
- Gender is created not only by biology, but by social practices.

CONTRIBUTIONS OF RADICAL ACTIVISTS

- The dangers of radioactive, toxic, and hazardous wastes to human health and reproduction have been exposed by citizen activists and regulations concerning disposal have been tightened.

- The siting of incinerators and landfills in poor and minority communities and Third World countries has been exposed as racist.
- The rapid clearcutting of tropical rainforests and northern hemisphere old growth forests by corporations on both public and private lands and the associated decimation of rare and endangered species have been brought to public awareness, and cutting in some areas has been curtailed.
- The slaughter of whales, dolphins, salmon, and other ocean species has been sharply criticized and in some cases curtailed or temporarily reduced.
- The dangers of pesticides and herbicides on foods and in water supplies and the availability of alternative systems of agriculture have been made visible.
- The viability of green parties as a source of political power has been recognized.
- The self-determination and power of indigenous peoples throughout the world to the right to control their own natural resources has become important.
- Direct, nonviolent action has become an acceptable and highly visible means of political protest.
- Alternative, nonpatriarchal forms of spirituality and alternative pathways within mainstream religions that view people as caretakers and/ or equal parts of nature rather than dominators are being adopted by more and more people.
- The need for ecological education and individual commitment to alternative lifestyles that reduce conspicuous consumption and recycle resources is making headway.

While radical ecology has achieved specific gains and visibility, it nonetheless has its own limitations and internal contradictions. Radical ecology lacks coherence as a theory and as a movement. Theoreticians are deeply divided as to underlying ethical, economic, social, and scientific assumptions. Some deep ecologists wish to focus on redefining the meaning of self, others on redefining science and cosmology, still others on the connections between spirituality and deep ecology. Social ecologists and deep ecologists are at odds as to whether the priority lies with challenging and redefining the dominant worldview as the mode for initiating transformation or whether the preeminent strategy lies in the pursuit of social justice, with each camp accusing the other of lack of sophistication. Some social ecologists disdain spiritual ecology as politically naive and as diverting energy away from social change, while many spiritual ecologists defend ritual as a way of focusing social actions. Ethically the camps are also in disagreement, with many deep ecologists and spiritual ecologists holding some form of ecocentric ethic, while social ecologists generally pursue a homocentric approach informed by ecological principles. Although the theoretical debates among proponents of radical ecology in general are often vituperative, they are equally incisive and healthy as a forum for clarification of assumptions and principles.

Similarly, green movements are divided along both theoretical and strategic lines. Green politics is fraught with disagreements between those who hold deep ecological and/or spiritual ecological assumptions and those who identify with social ecology and hold an ethic of social justice as the primary objective. Equally significant are the divisions between Greens who wish to pursue a practical real-world strategy of working with other political parties to achieve ecological goals and Greens who refuse to compromise fundamental movement principles and prefer to work outside the established political system. Ecofeminists are often critical of deep ecologists for their failure to recognize both biological and socially constructed differences, and divided among themselves as to basic strategies for change, with some pressing for spiritual, others for social approaches, and still others seeking to combine ritual with action. Similarly the sustainability movement is divided among those who primarily follow scientific/ecological principles in advocating policy and those who incorporate or subordinate scientific strategies to social justice strategies.

Radical environmental movements also differ in different parts of the world. In the First World, much energy is directed toward mitigating the effects of toxic pollutants (e.g. chloroflurocarbons, petroleum spills, PCBs, pesticides, and nuclear and hazardous wastes), preserving endangered species, saving wilderness, and promoting recycling. In the Second World, priorities are focused on controlling industrial threats to human health, particularly the effects of urban air and water pollution as well as nuclear contamination resulting from the Chernobyl accident. In the Third World a primary emphasis is on obtaining sufficient food, clean water, and adequate clothing for basic subsistence, developing appropriate technologies for cooking, heating, and farming, countering the effects of pesticide poisoning on human health, and preserving the lands of indigenous peoples.

Yet just as the environmental and human health problems facing the three worlds are interdependent, so radical movements are linked. When toxic substances and pharmaceuticals are banned in the First World, they are often dumped in Third World countries. Radical movements expose and protest against such practices. When rainforests are cut in Third World countries, destroying indigenous habitats, First World environmental groups organize consumer boycotts of timbers and hamburgers. When Second World activists organize environmental protests, they receive support and assistance from First World activists. International environmental conferences produce international networks of groups helping other groups.

Within the First, Second, and Third World radical ecology movements, theory and practice are linked, each informing and inseparable from the other. Divisions among proponents open new avenues for both synthesis and criticism. The movement as a whole is both dynamic and timely. New ideas and new strategies for change are continually evolving; the door is always open to new people with energy and enthusiasm.

I have organized the preceeding chapters around a framework that uses the concepts of ecology, production, reproduction, and consciousness in understanding both the ecological crisis and ways of overcoming it. I have analyzed the crisis a result of two contradictions, the first between production and ecology, the second between production and reproduction (see Introduction and Chapter 1). As these contradictions deepen, they push the world into greater ecological stress. The crisis could be relieved over the next several decades, however, through a global ecological revolution brought about by changes in production, reproduction, and consciousness that lead to ecological sustainability. Thus deep ecologists call for a transformation in consciousness from a mechanistic to an ecological worldview which transforms knowing, being, ethics, psychology, religion, and science, while spiritual ecologists focus on religion and ritual as ways of revering nature. Social ecologists call for a transformation in political economy based on new ecologically sustainable modes of production and new democratic modes of political reproduction.

Radical ecological movements attempt to resolve the contradictions that lead to the crisis through action. Green politics address the contradiction between production and reproduction, pressing for ways of reproducing human and nonhuman life that are compatible with ecosystem health and social justice. Ecofeminists press for gender equality and the subordination of production to the reproduction of life such that children will be born into societies that can provide adequate employment and security and have an ethic of nurturing both humans and nature. The sustainability movement focuses on the contradiction between ecology and production, devising ecologically-sustainable production technologies, restoring ecosystems, and promoting socially-just development programs.

Despite the accomplishments and vision of radical ecologists, however, most of the world's power is presently concentrated in economic systems and political institutions that bring about environmental deterioration. The trends that split rich from poor, whites from people of color, men from women, and humans from nature remain. Radical ecology itself stands outside the dominant political, economic, and scientific world order. Together its various strands and actions challenge the hegemony of the dominant order. Because environmental problems promise to be among the most critical issues facing the twenty-first century, environmentalists will play increasingly important roles in their resolution. Radical ecology and its movements will continue to challenge mainstream environmentalism and will remain on the cutting edge of social transformation, contributing thought and action to the search for a livable world.

NOTES

INTRODUCTION:
WHAT IS RADICAL ECOLOGY?

1. For this approach to self and society, I am indebted to my colleague Alan Miller's, *Gaia Connections: An Introduction to Ecology, Ecoethics, and Economics* (Savage, Md.: Roman and Littlefield, 1990), pp. 221–31, to Jerry W. Sanders who formulated and used it in a world order course at City College of New York, and to my students in Conservation and Resource Studies, University of California, Berkeley, who are quoted anonymously below.

CHAPTER 1
THE GLOBAL ECOLOGICAL CRISIS

1. "Planet of the Year: Endangered Earth," *Time*, January 2, 1989. Jane Kay, "Alaska's Eco-Disaster," *San Francisco Examiner*, August 27, 1989; "The Environment: A Higher Priority," *The New York Times*, July 2, 1989; Carolyn Merchant, "Gaia's Last Gasp," *Tikkun*, 5, no. 7 (March/April 1990): 66–9.

2. Associated Press, "Senators Unveil Legislative Plan to Combat Greenhouse Effect," San Francisco Chronicle, July 29, 1988.

3. "Fossil Fuels CO_2 Emissions: Three Countries Account for 50% in 1986" *CDIAC Communications*, Oak Ridge National Laboratory (Winter 1989).

4. Charles Petit, "Why the Earth's Climate is Changing Drastically," *San Francisco Chronicle*, August 8, 1988.

5. New York Times, "EPA Urges Drastic Action to Slow Greenhouse Effect," *San Francisco Chronicle*, March 14, 1989.

6. Michael D. Lemonick, "Deadly Danger in a Spray Can," *Time*, January 2, 1989, p. 42; Kara Swisher, "Refrigerators New CFC Issue," *Star Bulletin and Advertiser*, Honolulu, July 16, 1989, p. D–3.

7. Daniel Keith Conner and Robert O'Dell, "The Tightening Net of Marine Plastics Pollution," *Environment*, 30, no. 1 (January–February 1988): 17–20, 33–36; "The Dirty Seas," *Time*, August 1, 1988.

8. Norman Myers, ed., *Gaia: An Atlas of Planet Management* (Garden City, N. Y.: Anchor, 1984), p. 40.

9. Vandana Shiva, "Address," Fate and Hope of the Earth Conference, Managua, Nicaragua, June 1989.

10. *San Francisco Chronicle*, August 20, 1989.

11. "Putting the Heat on Japan," *Time*, July 10, 1989, pp. 50–2.

12. Peter Raven, "The Global Ecosystem in Crisis," MacArthur Foundation Occasional Paper (Chicago, Ill.: The John D. and Catherine T. MacArthur Foundation, Dec. 1987), p. 7; Rainforest Action Network, "An Emergency Call to Action for the Forests and Their Peoples."

13. Frederic P. Sutherland, *San Francisco Chronicle*, August 21, 1989; Jeff Pelline, "High Tech Moves into the Woods," Business Extra, *San Francisco Chronicle*, August 28, 1989.

14. Elliot Diringer, "U.S. Awash in Toxic Chemicals—And Fear of Them," *San Francisco Chronicle*, October 17, 1988; Diringer, "Science is Anything But Exact on Toxic Risks," *San Francisco Chronicle*, October 18, 1988; Diringer, "Prop 65 Begins to

Affect Products, Buying Habits," *San Francisco Chronicle*, October 20, 1988. (Proposition 65 is the Safe Drinking Water and Toxic Enforcement Act of 1986.)

15. Carolyn Merchant, *Ecological Revolutions: Nature, Gender, and Science in New England* (Chapel Hill: University of North Carolina Press, 1989).

16. James O'Connor, "Uneven and Combined Development and Ecological Crisis: A Theoretical Introduction," *Race and Class*, 30, no. 3 (January–March 1989): 1–11.

17. Samir Amin, *Eurocentrism* (New York: Monthly Review Press, 1989), pp. 71–8; Alfred Crosby, *Ecological Imperialism: The Biological Expansion of Europe, 900–1900* (New York: Cambridge University Press, 1986).

18. Alan Miller, *A Planet to Choose: Value Studies in Political Ecology* (New York: Pilgrim Press, 1978), pp. 23–35.

19. Dick Thompson, "The Greening of the U.S.S.R.," *Time*, January 2, 1989, pp. 68–9; Interview with Igor Izodorovich Altshuler and Ruben Artyomovich Mnatsakanyan, "The Changing Face of Environmentalism in the Soviet Union," *Environment*, 32, no. 2 (March 1990): 4–9, 26–30; Murray Feshbach and Ann Rubin, "Soviets Confront Ecological Disaster," *San Francisco Chronicle*, February 14, 1990; Alexei Yablokov, *et al*, "Russia: Gasping for Breath, Choking in Waste, Dying Young . . . and Cutting Down Siberia," *Washington Post*, August 18, 1991, p. C3.

20. Marlise Simons, "Rising Iron Curtain Exposes Haunting Veil of Polluted Air," *New York Times*, April 8, 1990, pp. 1, 8.

21. Marshall I. Goldman, "The Convergence of Environmental Disruption," *Science*, 170 (October 2, 1970): 37–42, quotation on p. 37; Gene E. Mumy, "Economic Systems and Environmental Quality: A Critique of the Current Debate," *Antipode*, 11, no. 2 (1979): 26–33.

22. William J. Keppler, quoted in *Pacific World*, No. 13 (November, 1989), p. 4.

23. Paul Ehrlich and Ann Ehrlich, *The Population Explosion* (New York: Simon and Schuster, 1990), pp. 14–17, quotation on p. 17; see also Paul Ehrlich, *The Population Bomb* (New York: Ballentine Books, 1968).

24. Ehrlich and Ehrlich, *Population Explosion*, pp. 205–10.

25. Ehrlich and Ehrlich, *Population Explosion*, pp. 214–6.

26. Thomas Malthus, *Essay on Population* [1798] (New York; Penguin, 1970).

27. David Harvey, "Population, Resources, and the Ideology of Science," *Economic Geography*, 50, no. 3 (July 1974): 256–77.

28. Barry Commoner, *Making Peace with the Planet* (New York: Pantheon, 1990), pp. 155–68, quotation on p. 168.

29. Herman Daly, *Steady State Economics* (San Francisco: W. H. Freeman, 1977), pp. 14–39, quotations on pp. 17, 24.

CHAPTER 2
SCIENCE AND WORLDVIEWS

1. Carolyn Merchant, *The Death of Nature: Women, Ecology, and the Scientific Revolution* (San Francisco: Harper and Row, 1980), pp. 1–6; Smohalla, quoted on p. 28.

2. Merchant, *Death of Nature*, pp. 51, 63, 67, 5.

3. Francis Bacon, "The Great Instauration (1620) in *Works*, ed. James Spedding, Robert Leslie Ellis, and Douglas Devon Heath, 14 vols. (London: Longman's Green, 1870), vol. 4, p. 20; Bacon, "The Masculine Birth of Time," ed. and trans. Benjamin Farrington in *The Philosophy of Francis Bacon* (Liverpool, Eng.: Liverpool University Press, 1964), p. 62; Bacon, "De Dignitate et Augmentis Scientiarum" (written 1623) in *Works*, vol. 4, pp. 287, 294.

4. Bacon, "Novum Organum," Part 2, in *Works*, vol. 4, pp. 247, 246; Bacon, "Valerius Terminus," in *Works*, vol. 3, pp. 217, 219; Bacon, "The Masculine Birth of Time," trans. Farrington, *Philosophy of Francis Bacon*, p. 62; Bacon, "The Great Instauration," *Works*, vol. 4, p. 29.

5. Merchant, *Death of Nature*, p. 171.

6. René Descartes, "Discourse on Method (1637)," Part 4, in E. S. Haldane and G. R. T. Ross, eds., *Philosophical Works of Descartes* 2 vols. (New York: Dover, 1955), vol. 1, p. 119.

7. Merchant, *Death of Nature*, pp. 187–88; Joseph Glanvill, *Plus Ultra* (1668) (Gainesville, Fla.: Scholar's Facsimile Reprints, 1958), quotations on pp. 9, 13, 56.

8. Robert Boyle, *Works*, ed. Thomas Birch (Hildesheim, W. Germany: Olms, 1965), vol. 1, p. 310.

9. Merchant, *Death of Nature*, p. 193.

10. Merchant, *Death of Nature*, pp. 227–8.

11. Merchant, *Death of Nature*, pp. 228–9.

12. Descartes, "Discourse on Method," in *Philosophical Works*, vol. 1, p. 85.

13. Merchant, *Death of Nature*, pp. 229–30.

14. Descartes, "Principia Philosophiae (1644)," in *Oeuvres*, ed. Charles Adam and Paul Tannery (Paris: Cerf, 1897–1913), principle 53, p. 93.

15. Descartes, "Discourse on Method," in *Philosophical Works*, 93, 87, 89, quotation on p. 92.

16. Thomas Hobbes, "De Cive" (written 1642) in *English Works* (reprint edition, Aalen, W. Germany: Scientia, 1966), vol. 2, p. xiv.

17. Hobbes, "Leviathan," in *English Works*, vol. 3, quotations from Chap. 4, pp. 18, 20; Chap. 3, p. 17.

18. Hobbes, "Leviathan," (1651) in *English Works*, vol. 3, Chap. 5, pp. 29, 30.

19. Martin Heidegger, *Der Satz vom Grund*, quoted in Hubert Dreyfus, *What Computers Can't Do* (New York: Harper and Row, 1972), p. 242, note 16.

20. Merchant, *Death of Nature*, p. 234.

21. Heidegger, *The Question Concerning Technology* (New York: Harper and Row, 1977), pp. 21, 23.

22. Merchant, *Death of Nature*, pp. 234–5.

23. Merchant, *Death of Nature*, pp. 275–6.

24. Merchant, *Death of Nature*, pp. 277–8.

25. Merchant, *Death of Nature*, p. 279.

26. Merchant, *Death of Nature*, pp. 290–1.

27. Merchant, *Death of Nature*, p. 288.

28. Merchant, *Death of Nature*, pp. 288–9.

CHAPTER 3
ENVIRONMENTAL ETHICS AND
POLITICAL CONFLICT

1. Aristotle, *Nichomachean Ethics*, in Richard McKeon, ed., *The Basic Works of Aristotle* (New York: Random House, 1941), 1095^a lines 13–15, p. 937.

2. Clifford Geertz, "Religion as a Cultural System," in *The Interpretation of Cultures*, (New York: Basic Books, 1973, pp. 87–125; Charles Taylor, "Neutrality in Political Science," in Alan Ryan, ed. *The Philosophy of Social Explanation* (London: Oxford University Press, 1973), pp. 139–70, esp. 144–6, 154–55.

3. J. Hector St John de Crèvecoeur, "What is an American," in Crèvecoeur, *Letters from an American Farmer* [1782] (New York: E.P. Dutton, 1957), p. 36.

4. Palmer v. Mulligan, 1805, quoted in Morton J. Horwitz, *The Transformation of American Law, 1780–1860* (Cambridge, Ma.: Harvard University Press, 1981), pp. 1–4, 35–7, quotation p. 3.

5. Edmund S. Morgan, *The Puritan Dilemma: The Story of John Winthrop* (Boston: Little Brown and Co., 1958), pp. 7–8, 28.

6. Lynn White Jr., "The Historical Roots of Our Ecologic Crisis," *Science*, 155, no. 3767 (March 10, 1967): 1203–1207.

7. John Winthrop, "Winthrop's Conclusions for the Plantation in New England," in *Old South Leaflets* (Boston, 1629), no 50, pp. 4–5.

8. John Quincy Adams, in *Congressional Globe*, 29, no. 1 (1846): 339–42. Adams omits the Biblical phrase "replenish the earth."

9. Thomas Hart Benton, in *Congressional Globe*, 29, no. 1 (1846), pp. 917–8." Benton

reverses the Biblical word ordering from "replenish the earth and subdue it" to "subdue and replenish the earth."

10. Reverend Dwinell, quoted in John Todd, *The Sunset Land or the Great Pacific Slope* (Boston: Lee and Shepard, 1870), p. 252.

11. Garrett Hardin, "The Tragedy of the Commons," *Science*, 162 (1968): 1243–8; Garrett Hardin and John Baden, eds., *Managing the Commons* (San Francisco: W. H. Freeman, 1977).

12. Thomas Hobbes, "The Philosophical Rudiments Concerning Government and Society," ("De Cive," 1647), in William Molesworth, ed., *English Works*, 11 vols. (reprint edition, Aalen, W. Germany: Scientia, 1966), vol. 2, quotations p. 11.

13. Hobbes, *Leviathan* [1651] in *English Works*, vol. 3, p. 145. On the transformation of the use of the commons, see Carolyn Merchant, *The Death of Nature: Women, Ecology, and the Scientific Revolution*, (San Francisco: Harper and Row, 1980), pp. 42–68, 209–13. Also Susan Jane Buck Cox, "No Tragedy on the Commons," *Environmental Ethics*, 7, no. 1 (1985): 49–61.

14. Hobbes, *Leviathan*, p. 158.

15. Hardin, "Tragedy of the Commons," in *Managing the Commons*, pp. 20–1, 26–28. Hardin argues against the principle of Adam Smith that "decisions reached individually will, in fact, be the best decisions for an entire society," since it would imply *laissez faire* population control methods. He does not, however, question Smith's fundamental assumption (*Wealth of Nations*, 1776) that under capitalism an individual who "intends only his own gain," is "led by an invisible hand to promote . . . the public interest." (Hardin, p. 19)

16. Garrett Hardin, *Promethean Ethics: Living with Death, Competition, and Triage* (Seattle: University of Washington Press, 1980). On the concept of triage, see also David H. Bennett, "Triage as a Species Preservation Strategy," *Environmental Ethics*, 8, no. 1 (Spring 1986): 47–58.

17. Hardin, "Living on a Lifeboat," in *Managing the Commons*, pp. 261–79.

18. Jeremy Bentham, *An Introduction to the Principles of Morals and Legislation* [1789] (London: W. Pickering, 1823), vol. I, p. 2. John Stuart Mill, *Utilitarianism* [1861] (Indianapolis, Bobbs Merrill, 1957), p. 10.

19. Bentham, *Introduction to the Principles of Morals*, pp. 2–3. "An action . . . may be said to be conformable to the principle of utility, or for shortness sake, to utility, (meaning with respect to the community at large) when the tendency it has to augment the happiness of the community is greater than any it has to diminish it" (p. 3). Mill, *Utilitarianism*, pp. 22–3.

20. Mill, *Utilitarianism*, quotations, p. 22. See also the following statements by Mill on the primacy of the good of the whole over that of the individual: "The happiness which forms the utilitarian standard of what is right in conduct is not the agent's own happiness but that of all concerned. As between his own happiness and that of

others, utilitarianism requires him to be as strictly impartial as a disinterested and benevolent spectator" (Mill, p. 22). "Utility would enjoin first, that laws and social arrangements would place the happiness or the interest of every individual as nearly as possible in harmony with the interest of the whole" (Mill, p. 22). "A direct impulse to promote the general good may be in every individual one of the habitual motives of action, and the sentiments connected therewith may fill a large and prominent place in every human being's sentient existence" (Mill, p. 23).

21. Mill, *Utilitarianism*, quotation, p. 10.

22. Mill, *Utilitarianism*, pp. 34–40, quotations pp. 34, 36, 40, 22.

23. René Dubos, "Conservation, Stewardship, and the Human Heart," *Audubon Magazine* (September 1972): 21–28; John Passmore, *Man's Responsibility for Nature* (New York: Scribner's, 1974), ch 2; Robin Attfield, *The Ethics of Environmental Concern* (New York: Columbia University Press, 1983).

24. Dubos, "Conservation, Stewardship, and the Human Heart," p. 27.

25. Merchant, *Death of Nature*, pp. 246–52.

26. Oliver Wendell Holmes in Diamond Glue Co. v. United States Glue Co., 187 U.S. 611, 616 (1903), quoted in Arthur McEvoy, "Toward an Interactive Theory of Nature and Culture," *Environmental Review*, 11, no. 4 (Winter 1987): 289–305, see p. 294.

27. Roderick Nash, *Wilderness and the American Mind* (New Haven: Yale University Press, 1977), pp. 161–81.

28. Tim Palmer, *Stanislaus: The Struggle for a River* (Berkeley: University of California Press, 1982), pp. 46–29, 64–76, quotations, pp. 53, 163; Palmer, *Endangered Rivers and the Conservation Movement* (Berkeley: University of California Press, 1986), pp. 125–28.

29. Peter Singer, *Animal Liberation: A New Ethics for our Treatment of Animals* (New York: Avon, 1975); Tom Regan, *All That Dwell Therein—Essays on Animal Rights and Environmental Ethics* (Berkeley: University of California Press, 1982). In *Utilitarianism*, Mill wrote, "The standard of morality . . . may accordingly be defined [as] 'the rules and precepts for human conduct', by the observance of which an existence such as has been described might be, to the greatest extent possible, secured to all mankind; and not to them only, but, so far as the nature of things admits, to the whole sentient creation" (p. 16).

30. Attfield, *The Ethics of Environmental Concern* (New York: Columbia University Press, 1983).

31. Mill, *Utilitarianism*, pp. 40, 34, 22, see discussion above. Leopold, *Sand County Almanac*, pp. 224–5.

32. Mill, *Utilitarianism*, quotation, p. 10; on education see p. 35. Leopold, *Sand County Almanac*, quotation pp. 224–5; on education see pp. 207–214.

33. Roderick Nash, "Do Rocks Have Rights?" *The Center Magazine* (November/December 1977), pp. 1–12, quotation, p. 10.

34. On holism see J. C. Smuts, *Holism and Evolution* (New York: Macmillan 1926). For a critique of holistic thinking see D. C. Phillips, *Holistic Thought in Social Science* (Stanford, Ca.: Stanford University Press, 1976).

35. Barry Commoner, *The Closing Circle: Nature, Man, and Technology* (New York: Bantam, 1972), pp. 29–35, 188. For a critique of Commoner's holism see Don Howard, "Commoner on Reductionism," *Environmental Ethics*, 1, no. 2 (Summer 1979): 159–76.

36. Commoner, *Closing Circle*, pp. 221–3.

37. John P. Briggs and F. David Peat, *Looking Glass Universe: The Emergence of Wholeness* (New York: Simon and Schuster, 1984), pp. 249–52.

38. Ilya Prigogine and Isabelle Stengers, *Order Out of Chaos: Man's New Dialogue with Nature* (New York: Bantam, 1984).

39. David Bohm, *Wholeness and the Implicate Order* (Boston: Routledge and Kegan Paul, 1980), pp. 1–26, 172–213.

40. Donald Worster, *Nature's Economy: The Roots of Ecology* (San Francisco: Sierra Club Books, 1977), pp. 329–30, 339–48. Peter Berg, *Figures of Regulation: Guides for Re-Balancing Society with the Biosphere* (San Francisco: Planet Drum Foundation, n. d.); Berg, ed., *Reinhabiting a Separate Country: A Bioregional Anthology of Northern California* (San Francisco: Planet Drum Foundation, 1978); Raymond Dasmann and Peter Berg, "Reinhabiting California," *The Ecologist*, 7, no. 10 (1980): 399–401; Raymond Dasmann, "Future Primitive: Ecosystem People versus Biosphere People," *Coevolution Quarterly* (Fall 1976): 26–31; Dasmann, "Biogeographical Provinces," *Coevolution Quarterly* (Fall 1976): 32–37. Fritjof Capra and Charlene Spretnak, *Green Politics: The Global Promise* (New York: E. P. Dutton, 1984).

41. On the problem of intrinsic value see J. Baird Callicott, "Intrinsic Value, Quantum Theory, and Environmental Ethics," *Environmental Ethics*, 7 (1985): 257–275.

42. Holmes Rolston III, "Is There an Ecological Ethic?" in *Philosophy Gone Wild: Essays in Environmental Ethics* (Buffalo: Prometheus Books, 1986, pp. 12–29), quotation on pp. 19–20.

43. J. Baird Callicott, *In Defense of the Land Ethic: Essays in Environmental Philosophy* (Albany: State of New York University Press, 1989), pp. 153–4, 165–6, quotation on p. 174.

44. Donald Worster, "Conservation and Environmentalist Movements in the U.S.: Comment on Nash and Hays," in Kendall E. Bailes, ed., *Environmental History: Critical Issues in Comparative Perspective* (Lanham: University Press of America, 1985), pp. 258–263, see p. 262.

45. Karen J. Warren, "Toward an Ecofeminist Ethic," *Studies in the Humanities* (Dec. 1988): 140–156; Jim Cheney, "Nature and the Theorizing of Difference," *Contempo-*

rary Philosophy, 13, no. 1 (1990): 1–14; Jim Cheney, "Eco-Feminism and Deep Ecology," *Environmental Ethics*, 9, no. 2 (Summer 1987): 115–145; Marti Kheel, "Ecofeminism and Deep Ecology: Reflections on Identity and Difference," in Carol Robb and Carl Casebolt, eds., *Covenant for a New Creation* (New York: Orbis Books, 1991); Michael E. Zimmerman, "Feminism, Deep Ecology, and Environmental Ethics," *Environmental Ethics*, 9, no. 1 (Spring 1987): 21–44; Warwick Fox, "The Deep Ecology-Ecofeminism Debate and its Parallels," *Environmental Ethics*, 11 (Spring 1989): 5–25.

CHAPTER 4
DEEP ECOLOGY

1. Arne Naess, "The Shallow and the Deep, Long-Range Ecology Movement," *Inquiry*, 16 (1972): 95–100.

2. Bill Devall, "The Deep Ecology Movement," *Natural Resources Journal*, 20 (1980): 299–322; Devall and Sessions, *Deep Ecology* (Salt Lake City: Peregrine Smith, 1985).

3. William R. Catton, Jr. and Riley E. Dunlap, "A New Ecological Paradigm for Post-Exuberant Sociology," *American Behavioral Scientist*, 20, no. 1 (September/October 1980): 15–47, quotation on p. 36.

4. Fritjof Capra, *The Tao of Physics* (Berkeley, Ca.: Shambala, 1975); Capra, *The Turning Point* (New York: Simon and Schuster, 1982).

5. Fritjof Capra, "Deep Ecology: A New Paradigm," *Earth Island Journal*, (Fall 1987): 27–30, quotations on pp. 29, 30.

6. David Bohm, *Wholeness and the Implicate Order* (Boston: Routledge and Kegan Paul, 1980), quotation on p. 195..

7. Ilya Prigogine, *Order Out of Chaos: Man's New Dialogue with Nature* (New York: Bantam, 1984).

8. James Gleick, *Chaos: The Making of a New Science* (New York: Viking, 1987), pp. 9–32. Edward Lorenz, "Predictability: Does the Flap of a Butterfly's Wings in Brazil Set Off a Tornado in Texas?" presented to the annual meeting of the American Association for the Advancement of Science in Washington, D. C. on December 29, 1972. Edward Lorenz, Crafoord Prize Lecture, *Tellus*, 1984, 36A, 98–110.

9. Charles Birch, "The Postmodern Challenge to Biology," in David Griffin, ed. *The Reenchantment of Science: Postmodern Proposals* (Albany, New York: State University of New York Press, 1988), pp. 1–46.

10. James Lovelock, *Gaia: A New Look at Life on Earth* (New York: Oxford University Press, 1979); Lovelock, *The Ages of Gaia: A Biography of Our Living Earth*, (New York: W. W. Norton, 1988).

11. Glennda Chui, "The Mother Earth Theory," *San Jose Mercury News*, March 8, 1988, pp. 1C, 2C; James Kirchner, "The Gaia Hypothesis: Can It Be Tested?" *Reviews of Geophysics*, 27, 2 (May 1989): 223–35.

12. J. Donald Hughes, "Gaia: An Ancient View of our Planet," *Environmental Review*, 6, no. 2 (1982); Norman Myers ed., *The Gaia Atlas of Planet Management* (New York: Doubleday Anchor, 1984); William Irwin Thompson, ed., *Gaia: A New Way of Knowing* (Great Barrington, Ma.: Lindisfarne Press); Michael Allaby, *A Guide to Gaia: A Survey of the New Science of Our Living Earth* (New York: Dutton, 1989). Joseph Lawrence, *Gaia: The Growth of an Idea* (New York: St. Martin's Press, 1990); Alan Miller, *Gaia Connections: An Introduction to Ecology, Ecoethics, and Economics* (Savage, Md.: Rowman & Littlefield, 1990).

13. Joseph Needham, *Science and Civilization in China* (Cambridge, Eng.: Cambridge University Press, 1956), vol 2; J. Baird Callicott and Roger T. Ames, *Nature in Asian Traditions of Thought* (Albany: State University of New York Press, 1989).

14. Lao Tzu, *The Tao-Teh King*, tran. C. Spurgeon Medhurst (Wheaton, Ill.: Theosophical Publishing House, 1972), chapters 1, 22.

15. Lao Tzu, *The Tao-Teh King*, chapter 51; Capra, *The Tao of Physics*.

16. Callicott and Ames, ed., *Nature in Asian Traditions of Thought*, p. 15.

17. George Bradford, *How Deep is Deep Ecology* (Hadley, Ma.: Times Change Press, 1989), quotations on pp. 3, 14.

18. Stephan Elkins, "The Politics of Mystical Ecology," *Telos*, no. 82 (Winter 1989–90): 52–70, quotation on p. 63.

19. Ariel Kay Salleh, "Deeper than Deep Ecology: The Eco-feminist Connection," *Environmental Ethics*, 6, no. 4 (Winter 1984): 340–5.

20. Ruth Bleier, *Science and Gender* (New York: Pergamon, 1984), pp. 193–9.

21. Marcus G. Raskin and Herbert J. Bernstein, *New Ways of Knowing: The Sciences, Society, and Reconstructive Knowledge* (Totowa, N. J.: Rowman and Littlefield, 1987), quotations on p. 268.

22. Bleier, *Science and Gender*, pp. 199–207; Evelyn Fox Keller, *Reflections on Gender and Science* (New Haven: Yale University Press, 1985), pp. 162–5; Evelyn Fox Keller, *A Feeling for the Organism: The Life and Work of Barbara McClintock* (New York: W. H. Freeman, 1983), pp. 99–102; Sandra Harding, *The Science Question in Feminism* (Ithaca, N. Y.: Cornell University Press, 1986).

CHAPTER 5
SPIRITUAL ECOLOGY

1. John Seed, Joanna Macy, Pat Fleming, Arne Naess, *Thinking Like a Mountain: Towards a Council of All Beings* (Philadelphia, Pa.: New Society Publishers, 1988), pp. 79–90, quotations on pp. 85–6, 87–8.

2. Seed, et al, *Thinking Like a Mountain*, pp. 41–3.

3. Joanna Rogers Macy, *Despair and Personal Power in the Nuclear Age* (Philadelphia: New Society Publishers, 1983), pp. 22–37, principles on pp. 22–23, quotation on p. 37; Macy, "In League with the Beings of the Future," *Creation* (March April 1989): 20–22.

4. Carol P. Christ, "Why Women Need the Goddess: Phenomenological, Psychological, and Political Reasons," in Charlene Spretnak, ed., *The Politics of Women's Spirituality: Essays on the Rise of Spiritual Power Within the Feminist Movement* (Garden City, N. Y.: 1982), pp. 71–86, quotation on p. 73; Monica Sjöö and Barbara Mor, *The Great Cosmic Mother: Rediscovering the Religion of the Earth* (San Francisco: Harper and Row, 1987).

5. William Anderson, *Green Man: The Archetype of our Oneness with the Earth* (San Francisco: HarperCollins, 1990) Harvey Stein, "The Green Man: Workshops for all Men about Life, the Warrior, and the Earth and its Creatures," flyer, fall 1991.

6. Elinor W. Gadon, *The Once and Future Goddess* (San Francisco: Harper and Row, 1989), pp. 369–77; Terry Allen Kupers, "Feminist Men," *Tikkun*, 5, no 4 (1990): 35–8; Robert Moore, "Interview," *Wingspan: Journal of the Male Spirit* (Spring 1990); 1, 10–12; Shepherd Bliss, "Thoughts on the Orpheus Myth," *Wingspan* (Spring 1990): quotation on p. 6. On the men's movement see also special issue of *Utne Reader* (April/May 1986).

7. Marija Gimbutas, *The Goddesses and Gods of Old Europe, 6500–3500 BC* (Berkeley: University of California Press, 1982).

8. Pamela Berger, *The Goddess Obscured: The Transformation of the Grain Protectress from Goddess to Saint* (Boston: Beacon Press, 1985).

9. Gadon, *The Once and Future Goddess*, pp. 236 ff., color plates 21 and 22; 31 and 32; 35 and 36.

10. Gadon, *The Once and Future Goddess*, pp. 242–8; color plate 10, black and white plates 119, 120.

11. Riane Eisler, *The Chalice and the Blade* (San Francisco: Harper and Row, 1988), pp. xvii, 185–203, 105.

12. Starhawk, *The Spiral Dance: A Rebirth of the Ancient Religion of the Great Goddess* (New York: Harper and Row, 1979); Z. Budapest, *The Holy Book of Women's Mysteries* (Oakland Ca., 1979); Gadon, *The Once and Future Goddess*, pp. 233–8. For examples of goddess inspired poetry, see Janine Canan, ed., *She Rises Like the Sun: Invocations of the Goddess by Contemporary American Women Poets* (Freedom, Ca.: Crossing Press, 1989). For examples of pagan poetry see Celeste Newbrough, *Pagan Psalms* (Berkeley, Ca.: Onecraft, 1982).

13. Janet Biehl, "Goddess Mythology in Ecological Politics," *New Politics* 2 (Winter 1989): 84–105, quotations on pp. 85, 86, 91.

14. Paula Gunn Allen, *The Sacred Hoop: Recovering the Feminine in American Indian*

Traditions (Boston: Beacon Press, 1984), quotation on p. 1; Paula Gunn Allen, ed., *Spider Woman's Granddaughters: Traditional Tales and Contemporary Writing by Native American Women* (New York: Fawcett Columbine, 1989).

15. Beverly Hungry Wolf, "The Ways of My Grandmothers," *Woman of Power: A Magazine of Feminism, Spirituality and Politics*, No. 14 (Summer 1989): 60–61, excerpted from Beverly Hungry Wolf, *The Ways of My Grandmothers* (New York: William Morrow, 1980).

16. For example see Richard K. Nelson, *Make Prayers to the Raven: A Koyuknon View of the Northern Forest* (Chicago: University of Chicago Press, 1983), pp. 225–37.

17. T. C. McLuhan, complier, *Touch the Earth: A Self-Portrait of Indian Existence* (New York: Simon and Schuster, 1971); W. C. Vanderwerth, ed., *Indian Oratory: Famous Speeches By Noted Indian Chieftans* (Norman, Ok., University of Oklahoma Press, 1971).

18. J. Baird Callicott, "American Indian Land Wisdom?: Sorting Out the Issues," in Callicott, *In Defense of the Land Ethic: Essays in Environmental Philosophy* (Albany, N. Y.: State University of New York Press, 1989), pp. 203–19, see p. 204.

19. Callicott, "American Indian Land Wisdom?" p. 219.

20. Frederick W. Krueger, "Christian Ecology: Building a New Environmental Coalition of the Twenty-first Century," *Ecology Center Newsletter*, Berkeley, Ca, 18, no. 12 (December 1989): 3–4, quotation on p. 4.

21. Eco-Justice Working Group, "Environmental Stewardship," (Washington, D. C.: General Board of Church and Society, adopted 1984).

22. Doron Amiran, "Ecology in Jewish Tradition," *Ecology Center Newsletter*, Berkeley, Ca, 18, no. 12 (December 1989): 8.

23. Ravan Farhadi, "Islam and Ecology as Taught by the Qur 'an," *Ecology Center Newsletter*, Berkeley, Ca., 18, no. 12 (December 1989): 6–7.

24. Matthew Fox, O. P., "Creation-Centered Spirituality From Hildegard of Bingen to Julian of Norwich: 300 Years of an Ecological Spirituality in the West," in Philip Joranson and Ken Butigan, eds., *Cry of the Environment: Rebuilding the Christian Creation Tradition* (Santa Fe, N. M.: Bear and Company, 1984), pp. 85–106, quotations on pp. 96–7, 98–9, 91, 93, 91, 92, 100.

25. Matthew Fox, "A Call for a Spiritual Renaissance," *Green Letter*, 5, no. 1 (Spring 1989): 4, 16–17.

26. John B. Cobb, Jr., "Ecology, Science, and Religion: Toward a Postmodern Worldview," in David Ray Griffin, ed., *The Reenchantment of Science: Postmodern Proposals* (Albany, N. Y.: State University of New York Press, 1988), pp. 99–113, esp. pp. 99, 107–8.

27. John B. Cobb, Jr and David Ray Griffin, *Process Theology* (Philadelphia: Westminister Press, 1976), pp. 14 (quotation), 23, 65–7, 76–9, 152–3. See also John B. Cobb,

Jr., "Process Theology and an Ecological Model," in Joranson and Butigan, ed., *Cry of the Environment*, pp. 329–336; Cobb, "Ecology, Ethics, and Theology," in H. E. Daly ed., *Economics, Ecology, Ethics: Essays Toward a Steady-State Economy* (San Francisco: W. H. Freeman, 1973), pp. 162–176; Charles Birch and John Cobb, Jr., *The Liberation of Life: From The Cell to the Community* (Cambridge: Cambridge University Press, 1981); Alfred North Whitehead, *Process and Reality*, ed. David Ray Griffin and Donald W. Sherburne (New York: Free Press, 1978); Conrad Bonifazi, *The Soul of the World: An Account of the Inwardness of Things* (Lanham, Md.: University Press of America, 1978).

28. Cobb and Griffin, *Process Theology*, p. 79, quotations p. 76, 155.

29. Jay McDaniel, "Physical Matter as Creative and Sentient," *Environmental Ethics*, 5, no. 4 (Winter 1983): 291–317; McDaniel, "Christian Spirituality as Openess to Fellow Creatures," *Environmental Ethics*, 8, no. 4 (Spring 1986): 33–46; McDaniel, *Of God and Pelicans: A Theology of Reverence for Life*. Louisville, Kt.: Westminster John Knox, 1989.

30. Susan Armstrong-Buck, "Whitehead's Metaphysical System as a Foundation for Environmental Ethics," *Environmental Ethics*, 8, no. 3 (Fall 1986): 241–259, quotations pp. 243 (from Whitehead's, *Process and Reality*, ed. Griffin and Sherburne, p. 18), p. 246.

CHAPTER 6
SOCIAL ECOLOGY

1. J. Baird Callicott and Frances Moore Lappé, "Marx Meets Muir: Toward a Synthesis of the Progressive Political and Ecological Visions," in P. Allen and D. Van Dusen, eds., *Global Perspectives on Agroecology and Sustainable Agricultural Systems: Proceedings of the Sixth International Scientific Conference of the International Federation of Organic Agricultural Movements* (Santa Cruz: University of California Agroecological Program, 1988), vol. 1, pp. 21–30, quotation on p. 21.

2. Roger Gottlieb, *History and Subjectivity: The Transformation of Marxist Theory* (Philadelphia, Pa.: Temple University Press, 1987); Gottlieb, *An Anthology of Western Marxism: From Lukács and Gramsci to Socialist-Feminism* (New York: Oxford University Press, 1989), pp. 1–25.

3. Howard Parsons, ed., *Marx and Engels on Ecology* (Westport, Ct.: Greenwood Press, 1977), p. 133.

4. Parsons, ed., *Marx and Engels on Ecology*, quotations on pp. 176, 172, 171.

5. Parsons, ed., *Marx and Engels on Ecology*, quotation on p. 178.

6. Parsons, ed., *Marx and Engels on Ecology*, quotation on p. 179.

7. Parsons, ed., *Marx and Engels on Ecology*, quotations on pp. 174, 183.

8. Parsons, ed., *Marx and Engels on Ecology*, quotation on p. 177.

9. Murray Bookchin, "The Concept of Social Ecology," *Coevolution Quarterly* (Winter 1981): 14–22, quotation on p. 17, Bookchin, *Ecology of Freedom* (Palo Alto, 1982).

10. Bookchin, "The Concept of Social Ecology," quotation on p. 20.

11. Bookchin, "The Concept of Social Ecology," quotation on p. 15.

12. O'Connor's second contradiction is similar to my first contradiction, that between ecology and production. In my framework, the second contradition is that between production and reproduction (see Introduction).

13. James O'Connor, "Capitalism, Nature, Socialism: A Theoretical Introduction," *Capitalism, Nature, Socialism* 1 (Fall 1988): 11–38.

14. Sean Swezey and Daniel Faber, "Disarticulated Accumulation, Argoexport, and Ecological Crisis in Nicaragua: The Case of Cotton," *Capitalism, Nature, Socialism* 1 (Fall 1988): 47–68.

15. Richard Levins and Richard Lewontin, "Dialectics," *The Dialectical Biologist* (Cambridge, Ma.: Harvard University Press, 1985), pp. 267–88, quotation on p. 268.

16. Levins and Lewontin, "Dialectics," quotation on p. 280.

CHAPTER 7
GREEN POLITICS

1. Brian Tokar, "Marketing the Environment," *Zeta Magazine*, (February 1990), pp. 15–21, quotation from Ruckelshaus on p. 17.

2. National Campaign Against Toxic Hazards, *The Citizens Toxics Protection Manual* (Boston, 1988); "From Poison to Prevention," *Toxic Times*, 2, no. 3 (Fall Winter 1989): 3–5.

3. Will Collette, "Organizing Toolbox," *Everyone's Backyard*, 7, no. 1 (Spring 1989): 4–5; Sanford J. Lewis, "Turning Industrial Polluters into Good Neighbors," *Whole Earth Review* (Spring 1990): 116–21.

4. Dick Russell, "Environmental Racism: Minority Communities and their Battle Against Toxics," *Amicus*, 11, no. 2 (Spring 1989): 22–32, quotations on pp. 22, 23. Cynthia Hamilton, "Women, Home, and Community: The Struggle in an Urban Environment," *Race, Poverty, and Environment Newsletter*, 1, no. 1 (April 1990): 3, 10–13.

5. Jesus Sanchez, "The Environment: Whose Movement?" *Green Letter*, 5, no. 1 (Spring 1989): 3–4, 14–16.

6. Hawley Truax, "Minorities at Risk," *Environmental Action* (January February 1990),

pp. 20–21. Charles Lee, "Toxic Wastes and Race in the United States: A National Report on the Racial and Socio-Economic Characteristics of Communities with Hazardous Wastes Sites (New York: United Church of Christ Commission for Racial Justice, 1987).

7. Russell, "Environmental Racism," p. 25; Sharon McCormick, "Jackson Wants Cleaner Air for Richmond," *San Francisco Chronicle*, April 2, 1990, A–10.

8. Russell, "Environmental Racism," pp. 24–6; Truax, "Minorities at Risk," pp. 20–1.

9. Truax, "Minorities at Risk," pp. 20–1.

10. Gail E. Chehak and Susan Shown Harjo, "Protection Quandry in Indian Country," *Environmental Action* (January February 1990): 21–2; Claude Engle, "Profiles: Environmental Action in Minority Communities," *Environmental Action* (January February 1990): 23–4.

11. Sanchez, "The Environment, Whose Movement," quotations on pp. 14, 15.

12. Carl Anthony, "Why African Americans Should be Environmentalists," *Earth Island Journal*, 5, no. 1 (Winter 1990): 43–44, quotations on pp. 43–44.

13. Philip Shabecoff, "Environmental Groups Faulted for Racism," *San Francisco Chronicle*, February 1, 1990; Sanchez, "The Environment, Whose Movement," quotations on p. 15.

14. Fritjof Capra and Charlene Spretnak, *Green Politics: The Global Promise* (New York: E. P. Dutton, 1984), pp. 3–56.

15. Phil Hill, "The Crisis of the Greens," *Socialist Politics*, No. 4 (Fall/Winter 1985): 8–25; Editors, "Ideological Conflict in the German Greens," *Green Perspectives: A Left Green Publication*, No. 13 (December 1988): 1–5.

16. Janet Biehl, "Western European Greens: Movement or Parliamentary Party?" *Green Perspectives: A Left Green Publication*, No. 19 (February 1990): 1–7.

17. Abby Peterson and Carolyn Merchant, "'Peace with the Earth': Women and the Environmental Movement in Sweden," *Women's Studies International Quarterly* 9, no. 5 (1986): 465–79, on p. 476.

18. David Orton, "Problems Facing the Green Movement in Canada and Nova Scotia," *Green Web* (Saltsprings, Nova Scotia), Bulletin 17, p. 1

19. Margo Adair and John Rensenbrink, "SPAKA: Democracy at Work," *Green Letter/Greener Times* (Autumn 1989): 3–5.

20. *Green Letter/Greener Times* (Autumn 1989): 26, 34, 15, 31.

21. "Greens Advance in Elections, *Green Letter* (Spring 1990): 28–9; George Raine, "Green Party Likely to Make Ballot," *San Francisco Examiner*, December 29, 1991; Howard Hawkins, "Left Green Network Holds First Conference," *Green Letter/Greener Times* (Autumn 1989): 50; Eric Chester, "Toward a Left Green Politics: The Iowa Conference," *Resist*, no. 217 (July/August 1989): 3, 7.

22. William Mueller, "Green Reds: Environmental Action in the Soviet Union," *Amicus*, 11, no. 3 (Summer 1989): 8–9; Michael Redclift, "Turning Nightmares into Dreams: The Green Movement in Eastern Europe," *The Ecologist*, 19, no. 5 (September, October 1989): 177–83.

23. Mueller, "Green Reds," p. 9; "Estonian Greens Establish Priorities," *Green Letter/ Greener Times* (Autumn 1989), p. 54.

24. Edward Abbey, *The Monkeywrench Gang* (New York: Avon, 1975), quotation on p. 44.

25. David Foreman and Bill Haywood, ed. *Ecodefense: A Field Guide to Monkeywrenching*, 2nd ed. (Tucson, Az.: Ned Ludd Books, 1987), pp. 10–17, quotation on p. 14.

26. "EF! Takes to the Trees," *Earth First!*, 9, no. 13 (September 22, 1989): 1, 4; "EF! Carries Mt. Graham to Washington," *Earth First!*, 10, no. 4 (March 20, 1990): 1, 3; "Log Ship Lockdown," *Earth First!*, 10, no. 3 (February 2, 1990): 1, 5; "Colorado EF! Hits Cowboys Again," *Earth First!* (February 2, 1990): 1.

27. Forum, "Only Man's Presence can Save Nature," *Harper's*, 280, no. 1679 (April 1990): 37–48, quotation on p. 44.

28. Miss Ann Thropy, *Earth First!* (May 1, 1987) and Bill Devall, Interview with Dave Foreman, in *Simply Living* (P. O. Box 2095, N. S. W. Australia) as quoted in George Bradford, "How Deep is Deep Ecology?" *The Fifth Estate* (1987): 3–30, quotations on p. 17. For Foreman's apology see Stephen Talbot, "Earth First!: What Next?" *Mother Jones* (November/December 1990): 47–9, 76–80, see p. 80.

29. Quoted in George Bradford, "How Deep is Deep Ecology?" *The Fifth Estate* (1987): 3–30, quotations on p. 17.

30. Bill Weinberg, "Social Ecology and Deep Ecology Meet," *Earth First!*, 10, no. 3 (February 2, 1990): 10. See also Steve Chase, ed., *Defending the Earth: A Dialogue Between Murray Bookchin and Dave Foreman* (Boston: South End Press, 1991).

31. "What Works: An Oral History of Five Greenpeace Campaigns," *Greenpeace*, 15, no. 1 (January/February 1990): 9–13.

32. *Earthday Wall Street Action Handbook* (New York, 1990), quotation on p. 12.

33. *Earthday Wall Street Action Handbook*, quotation on p. 2.

CHAPTER 8
ECOFEMINISM

1. Françoise d'Eaubonne, "Feminism or Death," in Elaine Marks and Isabelle de Courtivron, eds. *New French Feminisms: An Anthology* (Amherst: University of Massachusetts Press, 1980), pp. 64–7, but see especially p. 25; Françoise d'Eaubonne, *Le Féminisme ou la Mort* (Paris: Pierre Horay, 1974), pp. 213–52.

2. Ynestra King, "Toward an Ecological Feminism and a Feminist Ecology," in Joan Rothschild, ed., *Machina Ex Dea* (New York: Pergamon Press, 1983), pp. 118–29; Janet Biehl, "What is Social Ecofeminism?" *Green Perspectives*, 11 (October 1988).

3. Alison Jaggar, *Feminist Politics and Human Nature* (Totawa, N. J.: Roman and Allanheld, 1983); Karen Warren, "Feminism and Ecology: Making Connections," *Environmental Ethics*, vol. 9, no. 1 (1987): 3–10.

4. Karen Warren, "Toward an Ecofeminist Ethic," *Studies in the Humanities* (December 1988): 140–56, quotation on p. 151.

5. Karen Warren, "The Power and the Promise of Ecological Feminism," *Environmental Ethics*, 12, no. 2 (Summer 1990): 125–46.

6. Jaggar, *Feminist Politics and Human Nature*, pp. 27–47.

7. Simon de Beauvoir, *The Second Sex* [1949] (London: Penguin Books, 1972), pp. 95–6; Betty Friedan, *The Feminine Mystique* (New York: Dell, 1963), pp. 11–27, 326–63; King, "Toward an Ecological Feminism and a Feminist Ecology," pp. 121–2; Rachel Carson, *Silent Spring* (Boston: Houghton and Mifflin, 1962), pp. 1–37.

8. Barbara Holzman, "Women's Role in Environmental Organizations," manuscript in possession of the author, Berkeley, Ca.

9. Sherry Ortner, "Is Female to Male as Nature is to Culture?" in Michelle Rosaldo and Louise Lamphere, ed. *Women, Culture, and Society* (Stanford, Ca.: Stanford University Press, 1974), pp. 67–87.

10. Merlin Stone, *When God Was a Woman* (New York: Harcourt Brace Jovanovich, 1976); Carolyn Merchant, *The Death of Nature: Women, Ecology, and the Scientific Rvolution* (San Francisco: Harper and Row, 1980); Carolyn Merchant, "Earthcare: Women and the Environmental Movement," *Environment*, 23, no. 5 (June 1981): 6–13, 38–40.

11. Starhawk, *The Spiral Dance: A Rebirth of the Ancient Religion of the Great Goddess* (San Francisco: Harper and Row, 1979); Carol Gilligan, *In a Different Voice: Psychological Theory and Women's Development* (Cambridge, Ma.: Harvard University Press, 1982); Nel Noddings, *Caring: A Feminist Approach to Ethics and Moral Education* (Berkeley: University of California Press, 1984).

12. Ortner, "Is Female to Male as Nature is to Culture?" For a recent anthology of varieties of ecofeminism see Irene Diamond and Gloria Ornstein, eds., *Reweaving the World: The Emergence of Ecofeminism* (San Francisco: Sierra Club Books, 1990).

13. Dorothy Nelkin, "Nuclear Power as a Feminist Issue," *Environment*, vol. 23, no. 1 (1981): 14–20, 38–39.

14. Merchant, "Earthcare," quotation on p. 38.

15. Karen Stults, "Women Movers: Reflections on a Movement By Some of Its Leaders," *Everyone's Backyard*, vol. 7, no. 1 (Spring, 1989): 1; Ann Marie Capriotti-Hesketh, "Women and the Environmental Health Movement: Ecofeminism in

Action," Department of Biomedical and Environmental Health Sciences, University of California, Berkeley, Ca., unpublished manuscript in possession of the author.

16. Merchant, "Earthcare," p. 13.

17. Susan Prentice, "Taking Sides: What's wrong with Eco-Feminism?" *Women and Environments*, (Spring 1988): 9–10.

18. Janet Biehl, "What is Social Ecofeminism?" *Green Perspectives*, No. 11 (October 1988): 1–8, quotation on p. 7.

19. Janet Biehl, *Rethinking Ecofeminist Politics* (Boston: South End Press, 1991), pp. 1–7, 9–19.

20. Friedrich Engels, "Origin of the Family, Private Property, and the State," in *Selected Works* (New York: International Publishers, 1968), p. 455; Engels, *Dialectics of Nature*, ed. Clemens Dutt (New York: International Publishers, 1940), pp. 89–90.

21. Abby Peterson, "The Gender-Sex Dimension in Swedish Politics," *Acta Sociologica*, 27, no. 1 (1984): 3–17, quotation on p. 6.

22. Carolyn Merchant, *Ecological Revolutions: Nature, Gender, and Science in New England* (Chapel Hill: University of North Carolina Press, 1989), p. 14.

23. Irene Diamond, "Fertility as a Sound of Nature: Echoes of Anger and Celebration," Department of Political Science, University of Oregon, Eugene, Oregon, unpublished manuscript in possession of author, p. 14.

24. For examples see Merchant, "Earthcare," pp. 7–13, 38–40.

25. Vandana Shiva, *Staying Alive: Women, Ecology, and Development* (London: Zed Books, 1988), p. 76.

26. Shiva, *Staying Alive*, pp. 55–77.

27. John Farrell, "Agroforestry Systems," in Miguel Altieri, *Agroecology: The Scientific Basis of Alternative Agriculture* (Berkeley: Division of Biological Control, University of California, Berkeley, 1983), pp. 77–83.

28. Wangari Maathai, *The Green Belt Movement: Sharing the Approach and the Experience* (Nairobi, Kenya: Environment Liaison Centre International, 1988), pp. 5–24, quotation on p. 5.

29. Maathai, *Green Belt Movement*, pp. 9–30. See also Lori Ann Thrupp, "Women, Wood, and Work in Kenya and Beyond," *UNASYLVA* (FAO, Journal of Forestry), (Dec 1984): 37–43.

30. Sithembiso Nyoni, "Women, Environment, and Development in Zimbabwe," in *Women, Environment, Development Seminar Report* (London: Women's Environmental Network, 1989), pp. 25–7, quotation on p. 26.

31. Maloba, Robleto, Letelier, Castro, at Managua Conference, June 1989.

32. Chee Yoke Ling, "Women, Environment, Development: The Malaysian Experience," in Women's Environmental Network, *Women, Environment, Development Seminar Report* (London: Women's Environmental Network, 1989), pp. 23–4.

33. Jeanne Rhinelander, "Crusader in Krakow," *Worldwide News: World Women in Environment*, 8, no. 2 (March–April 1990): 1, 7; Interview with Soviet Environmentalist: Dr. Eugenia V. Afanasieva, *Worldwide News* (September–October 1989): 1, 5, quotations on p. 5.

34. "Women Meet in Moscow to Talk Environment," *Worldwide News: World Women in Environment* (November–December 1989), pp. 1–2.

35. Olga Uzhnurtsevaa speaking at the conference on "The Fate and Hope of the Earth," Managua, Nicaragua, June 1989.

CHAPTER 9
SUSTAINABLE DEVELOPMENT

1. Letter from Eric Holt to the Environmental Project on Central America (EPOCA), Friends of the Earth, San Francisco, September 28, 1989.

2. World Commission on Environment and Development, *Our Common Future* (Oxford, Eng.: Oxford University Press, 1987).

3. Letter from Eric Holt, September 1989.

4. Miguel Altieri, "Ecological Diversity and the Sustainability of California Agriculture," in *Sustainability of California Agriculture: A Symposium* (Davis: U.C. Sustainability of California Agriculture Research and Education Program, [1985]), pp 103–119, quotation p. 106; Gordon K. Douglass, "Sustainability of What? For Whom?" in *Sustainability of California Agriculture: A Symposium* (Davis: U.C. Sustainability of California Agriculture Research and Education Program, [1985]), pp. 29–47, quotation, p. 38; Miguel Altieri, James Davis, and Kate Burroughs, "Some Agroecological and Socioeconomic Features of Organic Farming in California: A Preliminary Study," *Biological Agriculture and Horticulture*, 1 (1983): 97–107; George E. Brown, Jr., "Stewardship in Agriculture," in Gordon K. Douglass, ed., *Agricultural Sustainability in a Changing World Order* (Boulder, Co.: Westview Press, 1984), pp. 147–158.

5. Bill Liebhardt, "Why Systems Research?" *Sustainable Agriculture News*, 1, no. 3 (Spring 1989): 1.

6. Douglass, "Sustainability of What?" p. 40.

7. Bill Mollison and David Holmgren, *Permaculture One: A Perennial Agriculture for Human Settlements* (Maryborough, Australia: Dominion Press-Hedges and Bell, 1978; Bill Mollison, *Permaculture Two: Practical Design for Town and Country in Permanent Agriculture* (Maryborough, Australia: Dominion Press-Hedges and Bell, 1979).

8. Wes Jackson, *New Roots for Agriculture*, 2nd ed. (Lincoln: University of Nebraska Press, 1985), pp. 133–48.

9. Richard L. Doutt, "Vice, Virtue, and the Vedalia," *Bulletin of the Entomological Society of America*, 4 (1958): 119–23; K. S. Hagen and J. M. Franz, "A History of Biological Control," *History of Entomology, Annual Reviews* (1973): 433–76, see pp. 433–35, 441–44; Richard L. Doutt, "A Tribute to Parasite Hunters," in Cynthia Westcott, ed., *Handbook on Biological Control of Insect Pests* (New York: Brooklyn Botanic Garden Record, Plants and Gardens, 1960), pp. 47–51, see p. 51; Paul Debach, *Biological Control By Natural Enemies* (London: Cambridge University Press, 1974), pp. 92–100.

10. F. Wilson and C. B. Huffaker, "The Philosophy, Scope and Importance of Biological Control, in C. B. Huffaker and P. S. Messenger, eds. *Theory and Practice of Biological Control* (New York: Academic Press, 1976), p. 4; R. F. Smith, J. L. Apple, and D. G. Bottrell, "The Origin of Integrated Pest Management Concepts for Agricultural Crops," in J. L. Apple and R. F. Smith, eds., *Integrated Pest Management* (New York: Plenum Press, 1976), p. 12.

11. Perkins, *Insects, Experts, and the Insecticide Crisis* (New York: Plenum Press, 1982), p. 184. For a critique of chemical controls by an advocate of biological control see Robert van den Bosh, *The Pesticide Conspiracy* [1978] (New York: Doubleday Anchor, 1980).

12. John J. Berger, *Restoring the Earth* (New York: Knopf, 1985), pp. 69–78.

13. On the philosophy of restoration see William R. Jordan, "Thoughts on Looking Back," *Restoration and Management Notes,* 1, no.. 3 (Winter 1983): 2; Jordan, "On Ecosystem Doctoring," *Restoration and Management Notes,* 1, no. 4 (Fall, 1983): 2; Carolyn Merchant, "Restoration and Reunion with Nature, *Restoration and Management Notes,* 4, no. 2 (Winter 1986): 68–70. On Aldo Leopold and restoration see "Looking Back: A Pioneering Restoration Project Turns Fifty," *Restoration and Management Notes,* 1, no. 3 (Winter 1983): 4–10. On the techniques of restoration see John Cairns, Jr., "Restoration, Reclamation, and Regeneration of Degraded or Destroyed Ecosystems," in Michael Soulé, ed., *Conservation Biology: The Science of Scarcity and Diversity* (Sunderland, Ma.: Sinauer, 1986), pp. 465–484. Restoration is not only used to reestablish natural areas such as parks and nature reserves, but also as mitigation in development. Thus as airport may expand by filling in a marsh to construct an airstrip. As mitigation for the construction, the developer must artificially reconstruct another marsh in the vicinity.

14. Peter Berg, "Bioregions," *Resurgence,* No. 98 (May June 1983): 19; Peter Berg and Raymond Dasmann, "Reinhabiting California," *Ecologist,* 7, no. 10 (1980): 399–401, see p. 399; Jim Dodge, "Living by Life: Some Bioregional Theory and Practice," *CoEvolution Quarterly* No. 32 (Winter 1981): 6–12, quotation on p. 7; James Parsons, "On 'Bioregionalism' and 'Watershed Consciousness,'" *The Professional Geographer,* 37, no. 1 (February 1985): 1–6.

15. Raymond Dasmann, "Biogeographical Provinces," *CoEvolution Quarterly* (Fall 1976): 32–7; Dasmann, "Future Primitive: Ecosystem People versus Biosphere People," *CoEvolution Quarterly*, (Fall 1976): 26–31.

16. Kirkpatrick Sale, *Dwellers in the Land: The Bioregional Vision* (Philadelphia: New Society Publishers, 1991), pp. 41–51.

17. Seth Zuckerman, "Living There," *Sierra* (March April, 1987): 61–7.

18. Sale, "Dwellers in the land," quotation on p. 28.

19. Walter Truett Anderson, "The Pitfalls of Bioregionalism," *Utne Reader*, (February/March 1986): 35–38, quotation on p. 37.

20. C. G. R. Chavasse and J. H. Johns, *The Forest World of New Zealand, Realm of Tane-mahuta* (Wellington, N. Z.: A. H. & A. W. Reed, 1975), p. 10.

21. "Background to Maruwhenua, Notes on the Address by Shane Jones," Manager Maruwhenua (Maori Secretariat), Ministry for the Environment, in *Global Environmental Issues and Sustainability*, Proceedings of a Seminar, Wellington, N. Z., (March 1989), p. 20.

22. S. C. Chin et al, *Logging Against the Natives of Sarawak* (Selangor, Malaysia: Institute of Social Analysis, 1989), pp. 1–30, 57–64; Heather Dalton, "Fighting for Their Lives," *Simply Living*, 3, 1 (1987), pp. 18–26; Jennie Dell, "Voices from the Forest," *Habitat Australia* 19, 1 (February 1991), pp. 16–19; Jayl Langub, "Some Aspects of the Life of the Penan," *The Sarawak Museum Journal*, 40, no. 61, new series (December 1989), pp. 168–84; Chee Yoke Ling, "Women, Environment, and Development: The Malaysian Experience," in *Women, Environment, Development Seminar Report* (London: The Women's Environmental Network, 1989), p. 24.

23. Susanna Hecht and Alexander Cockburn, *The Fate of the Forest: Developers, Destroyers, and Defenders of the Amazon* (London: Verso, 1989), pp. 161–87, 227–30.

24. Anonymous, "Rain Forest Goes Commercial," *San Francisco Chronicle*, April 30, 1990.

25. Herb Kawainui Kane, *Pele, Goddess of Hawai'i's Volcanoes* (Captain Cook, Hi.: The Kawainui Press, 1987), pp. 10–17.

26. Timothy Egan, "Energy Project Imperils a Rain Forest," *New York Times*, January 24, 1990, p. B8, quotation from Palikapu Dedman, president of the Pele Defense Fund.

27. Robert J. Mowris, "Energy Efficiency and Least-Cost Planning: The Best Way to Save Money and Reduce Energy Use in Hawaii" (San Francisco: Rainforest Action Network, 1990).

28. The World Commission on Environment and Development, *Our Common Future* (Oxford, Eng.: Oxford University Press, 1987), pp. ix-xv, 8–9.

29. WCED, *Our Common Future*, pp. 67–91.

30. *NGO News* (Fall 1987).

31. *Green Web* (Saltsprings, Nova Scotia, Canada), Bulletin 16, December, 1989 pp. 3, 7.

32. Lester Brown, "Picturing a Sustainable Society," *Elmwood Newsletter*, 6, no. 1 (Spring 1990): 1, 4, 10.

33. Lori Ann Thrupp, "The Political Economy of the Sustainable Development Crusade: From Elite Protectionism to Social Justice," presented at the 1990 Annual Meeting of the Association of American Geographers, Toronto, Canada, April, 1990.

INDEX

*Please refer as well to suggested
readings at the end of each chapter.*

Abalone Alliance, 179
Abbey, Edward, 173
Abortion, 31–32
Acid rain, 19, 163
Activism. *See* Ecofeminism, environ-
  mental actions of; Grassroots activism;
  Green politics; Indigenous peoples'
  movements; Minority activism
Adams, John Quincy, 66
Adler, Margot, 118
Afanasieva, Eugenia V., Dr., 208
Affinity groups, 119
Africa, 30, 183, 202–4
African Americans. *See* Minority activ-
  ism; Racial issues
African elephant, 22
Agriculture, 139, 197. *See also* Farmers;
  Horticultural societies; Industrialized
  agriculture; Sustainable agriculture
Agroforestry, 202
Air pollution, 18–19, 172, 207–8
Alaskan oil spill, 18
Allen, Paula Gunn, 120, 254n.14
Altieri, Miguel, 213
Amazon rainforest actions, 224–26
American Indians. *See* Native Americans
Anderson, Walter Truett, 221–22
Animals, 12f., 27; extinction or endan-
  germent of, 20–22, 112, 176–77; liber-
  ation of, 74, 247n.29

Animism, 24, 41–44, 48, 121
Anthony, Carl, 166
Anthropocentric ethic. *See* Homocentric
  ethic
Anti-establishment paradigm, 231t.
Anti-nuclear movement, 118–19, 172,
  176–7, 179, 184
Anti-toxics movement, 162–64, 184–85,
  193–94. *See also* Toxic wastes
Aral Sea, 27
Aristotle, 50–51, 61
Arminian doctrine, 63, 64t., 66
Armstrong-Buck, Susan, 128
Arrowsmith, William, 122
Art, 12f., 18, 116
Atomism, 49–50, 56–58, 65t., 68
Attfield, Robin, 64t., 72, 74
Australian United Tasmaninan group,
  167

Baby markets, 196, 199
Bacon, Francis, 46, 48
Beauvoir, Simone de, 189
Beings, Council of All, 111–12
Bentham, Jeremy, 64t., 70–71
Benton, Thomas Hart, 66
Berg, Peter, 218
Berger, Pamela, 116
Bernstein, Herbert, 106
Biehl, Janet, 119, 194

Big Baram River, 223
Big Ten. *See* Group of Ten
Biocentric ethic, 181
Biogeochemical cycles, 12f.
Biological control, 64t., 149–50, 215–16.
    See also Integrated Pest Management
    (IPM)
Biology: dialectical, 150–53; ecological
    model of, 97–98; and holism, 78; and
    human nature, 187t., 191–92, 231t.
Bioregionalism, 62, 78, 87, 145, 217,
    219t., 260n.14; critique of, 221–22;
    roots and sources of, 218, 248n.40
Bioregional paradigm, 220t.
Bioregions, defined, 218
Biospheric dependence, 218
Biospheric equality, 86, 87t.
Biota, 12f., 75; extinction or endanger-
    ment of, 20–22, 112, 177
Birch, Charles, 97
Birth: "death of -", 20; defects, 192; pre-
    mature, 27; rate of, 32, 34–35
Black Elk, 88, 121
Blacks. *See* Minority activism; Racial
    issues
Bleier, Ruth, 106
Bliss, Shepherd, 114
Bly, Robert, 114
Body, extensions of the, 37–38, 137
Bohm, David, 59, 77, 93–94
Bookchin, Murray, 64t., 73, 142–46, 194;
    relations with Earth First!, 175–76
Bourgeois, Louise, 116
Boyle, Robert, 47, 69
Bradford, George, 102–3, 175
Brazil, 205–6, 225–26
Brown, Lester, 230
Bruno, Giordano, 88
Bruntland, Gro Harlem, 227
Bruntland Report, 227–230
Budapest, Z., 118
Buddhism, 64t., 76, 87

Callicott, J. Baird, 79–80, 122, 133–34
Campbell, Joseph, 114
Canada, 169, 183, 229; Green parties in,
    169
Capital, 37–38, 58
Capital (Marx), 139–40
Capitalism, 7, 34; appropriation and
    expansionism of, 29, 147, 231t., 232;
    and colonialism, 24–25; contradictions
    of, 147, 240; ecofeminist critique of,
    186–87t., 196–98; and externalities,
    153, 197, 228; history and growth of,
    23–26, 44–45; *laissez faire*, 62–63, 69–
    70; Marxist analysis of, 136–37, 140–
    41; and the mechanistic worldview,
    48, 59; and patriarchy, 184, 196; and
    population, 33–34. *See also* Corpora-
    tions; Production
Capitalist ecological revolution, 200,
    243n.15. *See also* Colonial ecological
    revolution
Capra, Fritjof, 89, 93, 103–4
Carson, Rachel, 64t., 189
Cash crops, 185, 202
Castillo, Aurora, 164
Castro, Giselda, 205
Catton, William, 88–92, 103
Center economies, 25, 35
Center for Process Studies, 127
Center for the Study of the Postmodern
    World, 127
Change, 69, 151–52, 236
Chaos: mathematical theory of, 96–97;
    order out of, 95–96, 127; systems ap-
    proach, 64t.
Chemical poisons and pollution, 22, 27,
    69–70, 165, 204, 216. *See also* Toxic
    wastes
Chernobyl nuclear disaster, 27
Ch'i, 101
Chile, 205
Chipko movement, 183, 185, 200–202
Christ, Carol, 113
Christ, the cosmic, 126
Christianity, 122–23, 126, 139,
    252ns.20–21. *See also* Judeo–Christian
    tradition
Christo, Javacheff, 18
Cities: and bioregionalism, 221; with en-
    vironmental problems, 26–27, 140,
    172–73; urban restoration, 217
Citizens' action groups, 158, 162, 164,
    166, 190, 193, 228
Civil disobedience, 158, 177–79
Clamshell Alliance, 179
Class (social), 2–3, 7–8, 44; anti-class
    posture, 87t., 104

Clements, Frederic, 218
Clock, as metaphor, 48–50, 95
Cobb, John, 127–28
Colonial ecological revolution, 200, 243n.15; and capitalism, 23–24. *See also* Capitalist ecological revolution
Colonialism, 23–24, 196, 206; neo-, 25
Committees of Correspondence, 158, 169
Commodities, 4, 22–23. *See also* Capitalism; Consumption
Commoner, Barry, 34–37, 64t., 73
Commons, the, 67–68, 146n.13
Communication and media, 147, 158
Community Right to Know Law, 163
Compesino Development Centers, 213
Compesino to Compesino, 211, 213, 232
Compost Coven, 118
Comte, August, 58
Consciousness, 12, 86, 192, 235, 239. *See also* Cosmologies; Environmental paradigms; Worldviews
Consciousness-raising, 231t.
Consumption, 12f., 28. *See also* Capitalism; Commodities
Context: dependence upon, 65t., 77, 94, 107; independence from, 49, 51–52, 65t.
Contraceptives, 196
Contradictions: of capitalism (O'Connor), 147, 240, 254n.12; of production and ecology (Merchant), 10, 12f., 38–39, 187t., 212; of production and reproduction (Merchant), 10–11, 12f., 38–39, 187t., 239, 254n.12
"Convergence of Environmental Disruption, The" (Goldman), 28
Convergence thesis, 28–29
Copernican universe, 96
Corporations, 185, 228; funding environmental groups, 160, 161t. *See also* Capitalism
Corpuscular theories, 49
Cosmologies, 9, 12f., 42, 128; in transition, 58, 96
Council of All Beings, 111–12
Craighead, Meinrad, 116
Creation spirituality, 124–26, 252n.24
Crèvecoeur, Hector St. John de, 63
Cultural ecofeminism, 184, 187t., 190–93; critique of, 193–94
Czechoslovakia, 27

Daly, Herman, 37–38
Dams. *See* Water and hydraulic power
Darwinian evolution, 79, 89, 95–96
Dasmann, Raymond, 218
Davies, Robertson, 6
DDT, 204, 216
Death rates, 34–35
Debt burden of Third World countries, 35, 228
Decentralization, 87, 145, 194
De Cive (Hobbes), 53, 67
Deep ecology, 11, 62, 64t., 114; defined, 85–86; and the ecocentric ethic, 87–88; ecofeminist critique of, 104–5; new paradigms of, 85, 88–93, 90–91t., 90–92f.; political critique of, 102–4, 145, 175–76, 237; principles of, 86–88, 87t.; sources and antecedents, 88, 100–102
Demeter, 115
Democracy, 207
Demographic transition, 34–37
Dependency, economic, 25
Depletion of resources. *See* Extraction
Deptford Trilogy (Davies), 6
Derham, William, 64t., 72
Descartes, René, 47, 49–53, 56, 69
Devall, Bill, 86
Developed countries, 36f., 200. *See also* First World; Second World
Developing countries. *See* Third World
Dialectical Biologist, The (Levins and Lewontin), 150
Dialectics, 146, 187t.; in biology and science, 150–53; compared to mechanism, 151–52; ecofeminist, 198; Hegelian, 135–36, 142; Marxist, 135–39
Dialectics of Nature (Engels), 138–39
Diamond, Irene, 199
Diet for a Small Planet (Lappé), 133
Dionysus, 114
Direct action movement. *See* Nonviolent direct action movement
Discourse on Method (Descartes), 47, 50
Diversification (crop), 228
Diversity, principle of, 87–88, 129
Dodge, Jim, 218
Dolphins, campaign to save, 177
"Dominant Western Worldview" (DWW), 88–89, 90–91t.

Domination: of nature, 54–55, 88, 186t.; social, 142–43, 145, 187t., 194

"Do Rocks Have Rights?" (Nash), 76

Douglass, Gordon, 213–14

Dualism, 56, 65t., 69, 80, 105; ancient Greek, 139

Dubos, René, 64t., 72

Dunlap, Riley, 88–92

Dury, John, 47

Earth: as animate mother, 41–44, 48, 116, 121, 124, 173, 222, 248–49n.44; as center of cosmos, 58, 96; as inanimate, 41–42, 48–50, 55; planetary processes of, 98–99

Earthday 1990, 177

Earth First!: The Radical Environmental Journal, 173

Earth First!: criticized, 145, 175; direct actions, 174–75, 180–81; literature of, 118, 173–74; radical members, 86; and social ecologists, 175–76

Earth Island Institute, 166

Eastern Europe. *See* Second World

Eastern philosophy: Taoism, 100–102; use of metaphor in, 102

Eaubonne, Françoise d', 184

Eckhart, Meister, 125–26

Ecocentric ethic, 11, 59, 80–81; concept of intrinsic value, 78–80; described, 62, 64–65t., 74–76; difficulties with, 78; in ecology movements, 87–88, 170–71, 181, 214, 231; grounded in cosmos, 64t., 74; and holism, 76–78; and spirituality, 112. *See also* Egocentric ethic; Homocentric ethic

Ecodefence: A Field Guide to Monkeywrenching, 173–74

Ecofeminism, 11, 62, 183, 186–87t., 209, 257n.12; critique of deep ecology, 104–5, 107, 238; *ecofeminisme,* 184, 256n.1; environmental actions of, 119, 183–85, 191, 200–202, 257n.15; ethics, 185; spirituality of, 64t, 113, 115–20, 191. *See also* Cultural ecofeminism; Liberal ecofeminism; Social ecofeminism; Socialist ecofeminism

Eco-Justice Project, 123

Ecological core, 12f.

Ecological crisis: 1, 10, 17, 86, 92f., 146, 242n.1, ns.3–7, n.9 and n.12; elements of, 22–23; and population issues, 31–37; resulting from contradictions, 9–11. *See also* Biota, extinction or endangerment of; Chernobyl nuclear disaster; Global warming; Ozone depletion; Pollution; Rainforests, depletion of; Toxic wastes

Ecological revolution, 12f., 13, 14, 200, 239, 243n.15

Ecological scientific paradigm, 230, 231t., 232, 238

Ecological spirituality. *See* Spiritual ecology

Ecology: human, 8, 77; and Marxist theory, 135, 148f.; and production, 9, 148f., 239, 254n.12; science of, 8, 75, 78, 86; theoretical, 77, 247n.26. *See also* Environmental paradigms; Radical ecology

Economic crisis, 146, 148

Economic dependency, 25

Economic growth, 28; inherent to capitalism, 25, 29; sustainable, 228–29

Economic and Philosophical Manuscripts of 1844 (Marx), 137

Economy: energy base of, 45; steady-state, 37–38; subsistence, 44, 58; surplus, 24. *See also* Market; Political economy; Production

Education, 231; and ethics, 71, 75–76, 247n.32

Egg, The (newsletter), 123

Egocentric ethic, 59, 61, 63, 64–65t.; assumptions, 66–67, 245n.8, 245–46n.9; and *laissez faire* capitalism, 62–63, 69–70; Egocentric ethic(continued): and liberalism, 188; and mechanism, 68–69; problems of, 74. *See also* Ecocentric ethic; Homocentric ethic

Ehrlich, Paul and Ann, 30–32

Eisler, Riane, 117, 119

Eliade, Mircea, 114

Elkins, Stephan, 103–4

Elmwood Institute, 118

Emergence (painting by Hirsch), 117

Empiricism, 58

Employment, 134, 149

Empowerment, 111–12

"Endangered Earth, The" (Christo), 18

Endangered species, 20–22, 112, 177
Energy: conservation of, 226–27; exchanges, 12f., 77, 95; mechanical, 51; renewable vs. nonrenewable sources, 37–38, 45
Engels, Friedrich, 135, 137–39, 196
Environmental crisis. *See* Ecological crisis
Environmental Defense Fund, 159, 190
Environmental ethics, 61, 64–65t.; defined, 62; and religion, 6–7. *See also* Ecocentric ethic; Egocentric ethic; Homocentric ethic
Environmental history, 8
Environmentalism, 13–14, 93, 180–81; first wave of, 159, 161; and progressive politics, 132–34, 143; and racial issues, 166–67
Environmentalism,: *See also* Ecofeminism; Green politics; Group of Ten
Environmental paradigms, 85, 88–89, 92f.; "New Ecological Paradigm" (NEP), 88, 90–91t.
Environmental Project on Central America, 179
Environmental Protection Agency (EPA), 160
Environmental regulation, 70, 160, 163–64, 174, 189, 225–26
Epistemological assumption, 49, 53–54
Essay on Population (Malthus), 32
Ethics, 12f., 103; and education, 71, 75–76, 247n.32; and religion, 62; utilitarian, 64–65t., 70–71, 246n.19, 246–47n.20; whether sequentially developed, 75, 80. *See also* Ecocentric ethic; Egocentric ethic; Environmental ethics; Homocentric ethic
Evolution, 79, 89, 95–96
Exchange, 12f. *See also* Exportation
Expansionism, 2
Experimental Essays (Boyle), 47
Experimental science, 45–47
Explicate order, 94
Exportation, 21–22
Externalities, 153, 197, 228
Extinctions of species, 20–21, 112
Extraction of resources, 12f., 23–24, 68, 102; and ethics, 63; opposition to, 185, 186t.; "reserves" for, 225–26

Exuberance, age of, 88, 92f.

Facts: acceptance of, 106; and values, 75, 78–79
Farmers, 134, 202–4; organic, 62, 212; women as, 202–3
Farming. *See* Agriculture
Fascism, 77–78
Fellowship in Prayer, 123
Female principle, 115–17
Feminine Mystique, The (Friedan), 189
Feminism, 6, 184, 186–87t., 190; radical, 194–95; and reconstructive knowledge, 106–7; and spirituality, 64t., 113–14, 117–20. *See also* Ecofeminism
Fertility. *See* Birth, rate of
Feudalism, 23–24, 136
First contradiction (Merchant), 9, 12f., 38–39, 187t., 212, 254n.12. *See also* Ecology; Production
First International Conference on Women, Peace, and the Environment, 208
First World, 22, 27–29, 39; center economies, 25, 35; environmental concerns of, 185, 238; women of, 185
Fiscal Crisis of the State, The (O'Connor), 146
Food production, 35–36
Force, external, 56–57
Foreman, Dave, 173, 175–76
Forests: agroforestry, 202; conservation or restoration of, 87, 124, 159, 188, 201, 205, 217; denudation of old growth, 21–22, 45, 139, 174; depletion of rainforests, 21, 111, 237
Fox, Matthew, Father, 124–26
France, Ecologist Party of, 168
Francis of Assisi, Saint, 88
Friedan, Betty, 189
Friends Committee on Unity with Nature, 123
Friends of the Earth, 159; in Brazil, 205–6; in Malaysia, 206
Friends, Religious Society of, 123
Friends of the River, 190

Gabriella Women's Coalition (Philippines), 199
Gadon, Elinor, 116

Gaia, 99, 112, 116, 248n.44, 249–50n.12
"Gaia: An Ancient View of our Planet"
  (Hughes), 99
Gaia Connections (Miller), 241
Gaia hypothesis, 98–99
Gandhi, Indira, 32; Mahatma, 21
Gandhian nonviolence, 88, 179
Gatherer-hunters, 87
Geb (Egyptian deity), 114
Geertz, Clifford, 62
Gender relations, 142–43, 194–95, 236.
  See also Women
Genesis, Book of, 64t., 66, 72, 123
Geocentric earth, 58
Geothermal development, 226
Germany, Green party in, 167–68
Gibbs, Lois, 162, 192
Gimbutas, Marija, 115, 119
Glanvill, Joseph, 47
Glasnost, 26, 171
Global commons, 68; and biospheric de-
  pendence, 218
Global ecological crisis. *See* Ecological
  crisis
Global ecological revolution, 12f., 14,
  239, 243n.15
Global warming, 18–19, 163
God, 57, 64t., 113, 125–26; as mother or
  mother-father, 116, 125–26
Goddess: iconography and images of,
  115–17, 120, 201, 226; spirituality of,
  113–14, 117–21, 191, 251ns.4,and 12
Goddess Obscured, The (Berger), 116
Golden Rule, 64t., 71, 75
Goldman, Marshall I., 28
Gorbachev, Mikhail, 26–27
Government regulation, 26, 70, 160,
  163–64, 174, 189, 225–26
Grandmother Woodchuck, 120
Grassroots activism, 158, 162, 164, 166,
  190, 193, 228. *See also* Ecofeminism,
  environmental actions of; Indigenous
  peoples' movements; Minority ac-
  tivism
Great Self, 87
Greco-Roman deities, 115–16
Greenbelt movement, 183, 200, 203
Green Committees of Correspondence,
  158, 169
Greenhouse effect, 18–19, 163

Green Man, 114, 251n.5
Green Marxists. *See* Social ecology
Green parties: German, 167–68; international groups, 158, 168–71; platform,
  167–68
Greenpeace, 164, 176–77, 181
Green politics, 13–14, 62, 78, 93, 157,
  180–81, 237; civil disobedience in,
  158, 177–79; ethical grounds for, 159,
  170–71, 181; left, right and center,
  64t., 157–58, 179, 238; need for coalitions in, 166; and progressive politics,
  132–34, 143, 175–76; in the Second
  World, 171–73. *See also* Anti-toxics
  movement; Bioregionalism; Grassroots activism; Green parties; Group
  of Ten; Sustainable development
  movement
Green Program USA, 158, 169–71
Green Revolution (agricultural), 20, 200,
  203
Green spirituality. *See* Spiritual ecology
Green Web of Nova Scotia, 169
Griffin, David Ray, 127–28
Group of Ten, 157–59, 166, 189; funding sources, 160, 161t.; organizations
  listed, 159–60; policies and actions of,
  159–62, 180
Grünen, die (Germany), 167
Guminska, Maria, Dr., 207

Hardin, Garrett, 64t., 66–68
Harding, Sandra, 106–7
Hartlib, Samuel, 47
Hartshorne, Charles, 127
Harvey, David, 33
Hawksbill turtle, 21
Hazardous wastes.: *See* Toxic wastes
Health issues, 148f., 230
Hegel, Georg, 135–36, 142
Heidegger, Martin, 55, 88
Heliocentric cosmos, 58
Hetch Hetchy dam, 73
Hierarchy, 142–43, 145, 187t., 194
Hildegard of Bingen, 125–26
Hillman, James, 114
Hinds, Cathy, 193
Hinduism, 87, 100
Hirsch, Gila Yellin, 116–17
Hobbes, Thomas, 53, 56, 64t., 66–67

Holism, 59, 89, 93, 236, 248n.35; elements of, 65t., 76–78; and fascism, 77–78; relation of whole to parts, 65t., 76
Holmes, Oliver Wendell, 72
Holographic model of the brain, 97
Holomovement, 59, 94
Home (film), 122
Homocentric ethic, 61, 64–65t., 72; in green politics, 159, 171, 181; and Marxism, 73, 134, 153; problems of, 74; and utilitarianism, 62, 70–71. *See also* Ecocentric ethic; Egocentric ethic
Horticultural societies, 87, 137–38
Hughes, J. Donald, 99
Human beings: as managers or stewards, 55, 64t., 72, 74, 123–24, 247n.23; nature and distinguishing traits of, 137, 186–87t., 191–92, 194; relation to Nature, 88, 137; unity with nonhuman beings, 65t., 77, 86, 110–13. *See also* Nonhuman beings; Nonliving beings
"Human Exemptionalism Paradigm" (HEP), 89, 90–91t.
Hume, David, 53
Hungary, 172
Hungry Wolf, Beverly, 120–21
Huygens, Christiaan, 50
Hydraulic power. *See* Water and hydraulic power

Idealization, in mechanism, 51, 56
Identity, principle of, 49–51, 69
Implicate order, 77, 94
Inanna, 115
Incineration of wastes. *See under* Toxic wastes
India, Chipko movement in, 183, 185, 200–202
Indigenous cultures, 80
Indigenous peoples' movements, 62, 118–19, 165, 222; Amazonian standoff movement, 224–26; *Maruwhenua*, 222–23; Penan tribe, 223–24. *See also* Native Americans
Indigenous species, 204, 217
Industrialism, 1, 133–34; and pollution, 26, 28, 171, 206–7. *See also* Capitalism; Socialism

Industrialized agriculture, 213; agribusiness, 185; factory farms, 200–201
Industrial revolution, 24
Industrial scientific paradigm, 220t., 220–21
Institute of Culture and Creation Spirituality, 124
Institute for Social Ecology, 184
Integrated Pest Management (IPM), 150, 216. *See also* Biological control
International Development Association, 228
International Whaling Commission, 176
Intrinsic value, 64t., 75, 78–80, 93, 127–29, 248n.41
"Is the Earth a Living Organism?" (symposium), 99
"Is Female to Male as Nature is to Culture" (Ortner), 190
Islam, 124, 252n.23
Italy, Green party of, 168
I-thou relationship, 41, 86

Jackson, Jesse, 165
Jackson, Wes, 215
Jeffers, Robinson, 88
Judaism: ecological, 124, 252n.22; female spirit in, 116–17
Judeo-Christian tradition, 24, 64t., 66, 88. *See also* Christianity; Judaism
Julian of Norwich, 125–26
Jung, Carl, 114

Kabbalah, 116–17
Kapital, Das (Marx), 139–40
Keller, Evelyn Fox, 106
Kenya, 30, 183, 203
Keppler, William, 30
King, Ynestra, 184, 195, 245n.2
Knowledge: context-dependent, 65t., 77, 94, 107; context-independent, 49, 51–52, 65t.; reconstructive, 106–7; science as model for, 58
Koebele, Albert, 215
Kosher laws, 124
Kuhn, Thomas, 105

Labor, 59, 230; and industrial capitalism, 33–34, 133. *See also* Workers
Laissez faire capitalism, 62–63, 69–70

Lake Baikal, 27
Land: ethic, 75–76, 88; use of, 2–3; wisdom, 120–22
"Land Ethic, The" (Leopold), 75
Land Institute, 215
Language, 12f., 137; ordering sense data, 53
Lao Tzu, 100
Lappé, Frances Moore, 133–34
Latin America, 204–5
Laws. *See* Environmental Regulation
Left Greens, 64t.
Legislative politics, 9, 25–26, 158, 160
Leibniz, Gottfried Wilhelm, 50, 54, 88
Leopold, Aldo, 64t., 75–76, 88
Leviathan (Hobbes), 66–67
Levins, Richard, 150–52
Lewontin, Richard, 150–52
Liberal ecofeminism, 184, 186t., 188–90
Liberation theology, 125
Liebig, Justus, 139
"Lifeboat ethics", 67–68
Life, forms of. *See* Biota
Literature, 44, 59
Lithuanian Green Movement, 172–73
Livermore Action Group, 119, 179
"Living on a Lifeboat" (Hardin), 68
Living standards, and population, 34–35
Local autonomy, 87
Local ecology, 9, 14. *See also* Bioregions
Locke, John, 53, 64t.
Logging, 21–22; actions against, 180, 223–27, 261n.22
Logical positivism, 57–58, 105. *See also* Mechanistic worldview
Lorenz, Edward, 96–97
Lotka, A. J., 37
Love Canal, 162, 192–93
Lovelock, James, 98–99
Luke (16:2), 72

Maathai, Wangari, 203
McDaniel, Jay, 128
Machine: ideal or model, 51; as metaphor, 48–50, 54–55, 89
Macy, Joanna, 111
Mainstream environmentalism. *See* Group of Ten
Malaysia, 21, 206, 223
Maloba, Kathini, 203–4

Malthus, Thomas, 32, 64t.
Malthusianism, 32–33, 103, 145, 175, 231t.
Mao Zedong, 101
Maoism, 101
Maori people, 222–23
Market system, 24, 44–45, 58, 184
Marx, Karl, 59, 133–34; on nature, 135, 137–38, 137–41, 139–41; on population, 33–34
Marxism, 34, 141f., 253n.2; and contemporary social movements, 135, 148f., 149; dialectical, 135–37; and ecology, 134, 137–40, 142, 146; and the homocentric ethic, 73, 134, 153. *See also* Social ecology; Socialist ecology
Marxist feminism, 184, 186t.
Mary (Virgin), 116
Mass culture, 145
Mathematical formalism, 57–58, 69. *See also* Context, independence from
Mathematical Principles of Natural Philosophy (Newton), 56–57
Matriarchy, 117, 120
Matrix (affinity group), 119
Matter: animate and inanimate, 56–57, 64t., 76, 94; consisting of particles, 49–50, 56–58, 65t., 68; reality of, 136. *See also* Animism
Matthew (25:14), 72
Mechanistic worldview, 11, 49; assumptions of, 49–55, 65t.; compared to dialectical, 151–52; dualistic, 56, 69; and the egocentric ethic, 68–69; legitimating captialism, 48, 59; Marx and Engels on, 135. *See also* Experimental science; Logical positivism
Mechtild of Magdeburg, 125–26
Media and communication, 147, 158
Mendes, Chico, 224–25
Mendoza, Don Jose Jesus, 211–12
Men's movement, 114–15, 251n.6
Merchant, Carolyn, theory of contradictions, 9, 12f., 38–39, 187t., 212, 239
Metaphors: the butterfly, 97; in Eastern philosophy, 102; Gaia hypothesis as a, 99–100; human-machine, 48–50, 54, 89, 95
Metaphysics, 64t.

Methodological assumption, 49, 52–53
Miljöpartiet de gröna (Sweden), 168–69
Mill, John Stuart, 64t., 70–71
Miller, Alan, 241,n.1, 243n.18, 250n.12
Mind, 54, 97
Mining, 45, 172
Minority activism, 164–67. *See also* Racial issues; Third World
Miriam (Biblical), 117
Miscarriages, 27, 192, 204
Miss Ann Thropy, 175
"Missa Gaia, A Mass in Celebration of Mother Earth" (Winter), 99
Momentum (mv), law of, 50–51, 56
Money, 44
Monkeywrench Gang, The (Abbey), 173
Monkeywrenching, 173–74
Monoculture, 145, 214
Moore, Robert, 114
Moral considerability, 75. *See also* Intrinsic value
Mosaic decalogue, 71, 75
Moslem religion, 124, 252n.23
Mother Earth, 41–44, 48, 116, 121, 124–25, 173, 222
Mothers of East Los Angeles (MELA), 164
Motion, discontinuous, 94
Muir, John, 88, 159
Myth, 9. *See also* Consciousness

Naess, Arne, 85–87, 104
Napier, John, 54
Nash, Roderick, 76
National Toxics Campaign, 22, 162–63, 179, 193
Native Americans, 43, 64t., 76, 99, 193, 200; land ethic of, 80, 120–22, 248–49n.44
Native Americans for a Clean Environment (NACE), 165
Native cultures, 80
Native species, 217
Natural Resources Defense Council, 159
Nature, 12f., 187, 194, 198; as alive and active, 11, 149, 187t., 196, 253n.29; as a commodity, 4; "death of -", 56, 58, 244n.1; human relation to, 54–55, 88, 137, 186t.; natural disasters, 188; rational ordering of, 49–51, 69; as so-

cially constructed, 194; women linked to, 59, 184, 190–91, 197. *See also* Earth; Nonhuman beings; Nonliving beings
Nature philosophers, German, 59
Nature spirituality, 110–11, 113–19; the Council of All Beings, 111–13; and mainstream religions, 122–24, 126–29; Native American, 120–22. *See also* Creation spirituality; Ecofeminism, spirituality of; Goddess; Mother Earth; Process, in theology
Needham, Joseph, 88
"New Ecological Paradigm" (NEP), 88, 90–91t.
New Testament, 72. *See also* Christianity
Newton, Isaac, 50, 57, 69
Newtonian science, 55–58, 94–95
New Ways of Knowing (Raskin and Bernstein), 106
New Zealand Values Party, 167
Nicaragua, 149–50, 204–5, 213; Environmental Movement of, 204–5
Nichomachean Ethics (Aristotle), 61
Nonhuman beings: as machines, 69; preliterate cultures toward, 99, 121, 142, 252n.16; rights and intrinsic value of, 64t., 76, 78, 128; unity with humans, 65t., 77, 86, 110–13
Nonlinear relationships, 77, 96, 99
Nonliving beings, 64t., 76, 94, 99, 127–28
Nonviolent direct action movement, 88, 118–19, 177–79, 237; successes and failures, 180. *See also* Anti-nuclear movement; Anti-toxics movements
North American Conference on Christianity and Ecology, 123
Nova Scotia, Green Web of, 169
Nuclear power: Chernobyl, 27; movement against, 118–19, 172, 176–7, 179, 184; radioactive wastes, 113, 181, 236

O'Connor, James, on contradictions of capitalism, 146–47
Old Testament, 71, 75. *See also* Judeo-Christian tradition

Once and Future Goddess, The (Gadon), 116
Ontological assumption, 49–50
Opticks (Newton), 57
Optics, 57, 77
Order: implicate-explicate, 77, 94; out of chaos, 95–96; rationalizing nature to attain, 49–51, 69; and sense data, 53
Organic farmers, 62, 212
Organicism (contemporary). *See* Holism
Organic worldview, 11, 42–44, 48, 59; conceptual elements, 56–57; earth as animate being, 41, 48, 116, 121, 124, 173, 222, 248–49n.44; restrained exploitation, 45. *See also* Holism
Origin of the Family, Private Property, and the State (Engels), 196
Orpheus, 114
Ortner, Sherry, 190
Osiris, 115
Our Common Future. See Bruntland Report
"Overdevelopment", 23, 189
Overproduction, 148f.
Ozone depletion, 19, 163

Pacific Stock Exchange Action, 178–79
Paganism, 118, 251n.12. *See also* Greco-Roman deities; Wicca
Pan, 114
Pan-African Women's Trade Union, 204
Pan Robin of the Green, 114
Paraguay proposal, 229
Parklands, 216–17, 232
Parmenides of Elea, 50
Particulate matter, 49–50, 56–58, 65t., 68
Partnership ethic, 117, 188
Parts: emphasis on, 49, 52–53; relation to whole, 65t., 76
Pascal, Blaise, 54
Patriarchy, 5, 142–43, 187t., 191, 194, 257n.10; and capitalism, 184, 196
Pele Defense Fund, 226
Penan tribe, 223–24
Perestroika, 171
Peripheral economies, 25, 35
Permaculture, 214–15
Perry, Ted, 122
Persephone, 115

Pesticides, 149–50, 204
Peterson, Abby, 197
Philippines, 199
Philosophiae naturalis principia mathematica (Newton), 56–57
Philosophy, 12f., 59; Presocratic, 88; process, 64t., 127
Physics, 58, 93–94
Pinchot, Gifford, 64t., 159
Place, politics of, 220–21
Planet Drum Foundation, 220
Planned obsolescence or longevity, 38
Plants, 12f., 20–21. *See also* Forests; Rainforests
Plastic wastes, 19–20. *See also* Toxic wastes
Plato, 50–51, 57
Plumwood, Val, 195
Plus Ultra (Glanvill), 47
Poland, 27, 207–8; environmental movements in, 172
Political economy, 23–26, 39. *See also* Capitalism; Socialism
Politics. *See* Grassroots activism; Green politics; Legislative politics; Social movements
Pollution, 87; air, 18–19, 172, 207–8; industrial, 26, 28, 171, 206–7; water, 19–20, 76. *See also* Toxic wastes
Population, 17, 25, 29, 31–34; control of, 68, 230, 246n.15; the demographic transition, 34–37; issues regarding, 145, 153, 199; rate of growth, 30–31, 31f., 36f.; theories of, 31–37
"Post-exuberant age", 89, 92f., 249n.9
Post-Marxism, 73
Postmodernism, 127
Power, and science, 51, 55
Predictability, 97. *See also* Chaos
Prentice, Susan, 193
Pribram, Karl, 97
Prigogine, Ilya, 77, 94–96
Process: defined, 127; in philosophy, 64t., 127; primacy of, 65t., 77, 88; in science, 93–94, 102, 142; in theology, 126–29, 252–53n.27
Production: and ecology, 17, 22–23, 186–87t., 207; and ecology (contradictions), 10, 12f., 38–39, 187t., 212; interactions and relations of, 58, 147,

148f.; and reproduction, 13, 29–30, 197–99, 240; and reproduction (contradictions), 10–11, 12f., 38–39, 187t., 239; systems or modes of, 9, 23, 29, 44, 140–42, 148f. *See also* Exchange; Extraction; Political economy
Progress, ideology of, 88–89
Progressive ecology. *See* Social ecology
Protestant ethic, 63, 64t., 66
Ptolemaic cosmos, 58, 96
Public attitudes, 18

Quakers, 123
Quantum theory, 80, 93–94
Quarks, 58

Racial issues, 59, 164–66, 177
Radical ecology, 239; contributions of, 235–237; critique of, 237–39; defined, 1–2, 9–10; and the environmental movement, 13–14. *See also* Deep ecology; Ecofeminism; Social ecology; Spiritual ecology
Radical feminism, 194–95
Radioactive wastes, 113, 181, 236. *See also* Toxic wastes
Rainbow Warrior, 177
Rainforests: actions to defend, 223–27, 261n.22; depletion of, 21, 111, 237
RARE II Report, 174
Raskin, Marcus, 106
Ray, John, 64t., 72
Realos vs. fundis, 168
Reconstructive science, 105–7
Recycling, 123; Marx on, 140
Redevelopment, 172
Red greens. *See* Social ecology
Red Sea, The (painting by Swartz), 117
Redwood Summer, 180
Regan, Tom, 74
Regulations, governmental, 26, 70, 160, 163–64, 174, 189, 225–26
Religion, 9, 12f., 57, 62; alternative spirituality, 110–11, 113, 129; concept of stewardship, 64t., 72, 74, 123–24; creation-centered, 124–26; doctrine of individual salvation, 63, 64t., 66; and environmental ethics, 6, 64–65t., 76, 78, 111–12; indigenous or Native American, 24, 120–22; mainstream,

122–24, 126–29; process theology, 126–29. *See also* Buddhism; Christianity; Hinduism; Islam; Judaism; Taoism; Wicca
Renaissance worldview. *See* Organic worldview
Reproduction: biological, 10, 12f., 22, 187t.; and ecofeminist theory, 195–96; interaction with production, 10–11, 29–30, 38–39, 197–99, 239–40; linking women with nature, 190–91; social or cultural systems of, 9, 11, 22, 58, 187t.; technologies of, 196, 199, 201; threats to human, 27, 32, 162, 185, 192, 196, 198; threats to species, 20–22, 112. *See also* Population
Reproductive freedom, 187
Resource Conservation and Recovery Act, 163
Resource extraction, 12f., 23–24, 68, 102; and ethics, 63, 229–30; opposition to, 185, 186t.
Resource planning, 148f.
Restoration ecology, 64t., 216–17, 260n.13
Rethinking Ecofeminist Politics (Biehl), 195
Ricardo, David, 33
River Thames, 140
Roadless Area Review and Evaluation (1977–78), 174
Robleto, Maria Luisa, 204
Rolston, Holmes, III, 79
Romanticism, 59
Romantic love, 194
Roybal-Allard, Lucille, 164
Ruckelshaus, William, 161

Sale, Kirkpatrick, 220t., 220–21
Salleh, Ariel Kay, 104, 195
Sand County Almanac, A (Leopold), 75
Sandinistas, 150
San Francisco Bay, 70, 190
Santayana, George, 88
Sarawak, 223–24
Save the Bay Association, 190
Science, 9, 12f.; basis for ethics or values, 64–65t., 75, 78–79; dialectics of, 151–53; of ecology, 9, 75, 78, 86; experimental, 45–47; and power, 51, 55; process theory in, 93–94, 102, 142; so-

cial construction of, 104–08, 105–8, 151–52, 236, 250n.22; and worldviews, 41, 51

Scientific ecological paradigm, 230, 231t., 232, 238

Seal hunting, 176

Seattle, Chief, 121–22

Second contradiction (Merchant), 10–11, 12f., 38–39, 187t., 239. *See also* Production; Reproduction

Second Sex, The (de Beauvoir), 189

Second World, 22, 26–27, 39; environmental problems of, 27–29, 243ns.19–20; former Soviet Union, 171; green politics in, 171–73; women of, 207–9

Seed, John, 111

Self: as ethical ground, 62–63; new psychology of, 86–87; and society, 2–8

Sense data, selection of, 49, 53–54

Sensuousness, 137

Sentience, as ground for ethics, 70–71

Sessions, George, 86

Sex. *See* Birth; Gender Relations; Reproduction

"Shallow and the Deep, Long-Range Ecology Movement, The" (Naess), 85, 87

Shekinah, 116–17

Shelford, Victor, 218

Shepard, Paul, 88

Shintoism, 76

Shiva, Vandana, 20, 201–2, 242n.9, 258n.25

Sierra Club, 159, 167

Silent Spring, The (Carson), 189

Singer, Peter, 64t., 74

Sithembiso, Nyoni, 203

Smith, Adam, 64t.

Smith, Henry A., M.D., 122

Smohalla, 43, 121

Smuts, Jan Christiaan, 59

Snyder, Gary, 88

Social class, 2–3, 7, 44; anti-class posture, 87t., 104

Social construction: of human nature, 137, 194; of nature, 194; of science, 104–08, 105–8, 151–52, 236

Social contract, 67

Social ecofeminism, 64t., 184, 194–95

Social ecology, 9, 107, 132–34, 153–54, 239; anarchist, 142–45; critique of, 153, 237–38; and Earth First!ers, 175–76; and Marxist theory, 134–42, 146. *See also* Socialist ecology

Social good, as ethical ground, 62, 64t., 71

Socialism, 27, 171; and externalities, 153; state, 149. *See also* Marxism

Socialist ecofeminism, 184, 187t., 195–96; and production, 197–98; and reproduction, 198–200

Socialist ecology, 134, 148f., 198; theory of, 146–47

Socialization, 4–6, 12f.

Social justice, 62, 231t., 238

Social movements. *See* Nonviolent direct action movement; Feminism; Environmentalism; Indigenous peoples' movements; Minority activism

Social structure, 87; class, 2–3, 7, 44; hierarchy and domination in, 142–43, 145, 187t., 194

Society: self in, 1–4; versus self, 7–8. *See also* Socialization

Soil, 20, 212

Solar energy, 37, 43

Soviet Union, former, 171. *See also* Second World

Space, void, 57

Species. *See* Biota

Spinoza, Baruch, 88

Spiritual ecology, 12, 78, 87, 110, 129, 238–39; and creation

spirituality, 124–26; criticized, 129, 238; defined, 111–12; green, 64t., 86; and the greens, 64t., 86, 145; and stewardship, 64t., 72, 74, 123–24. *See also* Ecofeminism, spirituality of

Spotted Owl, 22

Standing Bear, Luther Chief, 88, 121

Stand-off movement, 224–25

Stanislaus River, 73–74

Starhawk, 118, 251n.12

State, 12f., 134. *See also* Legislative politics; Political economy

State socialism, 149, 207

Steady-state economics, 37–38

Sterilization, 32

Stewardship concept, 64t., 72, 74, 123–24, 247n.23

Stillbirths, 192
Structure of Scientific Revolutions (Kuhn), 105
Subject-object relationship, 78, 80
Suisuin Marsh, 70
Sun Dance camps, 121
Susan B. Anthony Coven, 118
Sustainable agriculture, 64t., 212, 259n.4; crop diversification, 228; definition and principles of, 213–14, 230; organic farming, 62, 212; permaculture, 214–15
Sustainable development movement, 62, 142, 198–99, 203–04, 232–33, 240; global, 227–30, 232, 239; sectors of, 230, 231t., 238–39
Suzuki, Daisetz, 88
Swartz, Beth Ames, 117
Sweden: green politics in, 168–69, 183
Symbiosis, 87
Synergy, 76, 217
Systems: approach and methodology of, 93, 103–4, 214, 231; bio-, 129; chaotic, 64t.; closed and open, 77, 95–96; ecological, 76, 144

Tammuz, 115
Taoism, 100–102
Tao of Physics, The (Capra), 89
Tao Te Ching (The Way), 100
Taylor, Charles, 62
Technology, 9, 88, 102; belief in, 48, 89, 230, 231t., 245n.21; natural restrictions on, 38. *See also* Production
Ten Commandments, 71, 75
Terra Mater, 116
Thames River, 140
Theology: liberation, 125; process, 126–29
Thermodynamics, laws of, 38, 94–96
Third World, 22, 39, 167, 200; debt burden, 35, 228; environmental concerns, 185, 238–39; extraction of raw materials, 8; and population growth rates, 36f.; women of, 185, 199–207. *See also* Peripheral economies
Thoreau, Henry David, 88
Thrupp, Lori Ann, 230–32
Tobias, Michael, 86
Toolmaking, 137

Torah, 124
Toxic chemicals. *See* Toxic wastes
Toxic Times, 162
Toxic wastes: actions regarding, 162–64, 184–85, 193–94; in breast milk, 205; chemical, 22, 26–27, 69–70, 204, 216, 237; disposal or incineration of, 140, 148f., 164–65, 168, 172, 177, 183, 237; hazardous, 162, 236; plastic, 19–20; radioactive, 113, 181, 236
"Toxic Wastes and Race in the United States", 164
Trade, 24, 44–45
"Tragedy of the Commons" (Hardin), 66–68
Transcendentalism, 59
Tree-embracing. *See* Chipko movement
Triage, policy of, 68, 246n.16
Tropical forests. *See* Rainforests
Tucker, Cora, 166
Turning Point, The (Capra), 89
Turtle, Hawksbill, 21

"Underdeveloped" world. *See* Third World
"Underdevelopment", 23
Underproduction, 148f.
Unemployment, 25
United Nations Biosphere Reserve, 224
United States: Green parties, 158, 169–71
United Tasmaninan Group (Australia), 167
"Unity of differences", 142
Urban restoration, 217. *See also* Cities
Utilitarianism, 70–71, 246n.19, 246–47n.20
Uzhnurtesevaa, Olga, 208–9

Value: instrumental, 78, 128; intrinsic, 64t., 75, 78–80, 93, 127–29, 248n.41
Values: and scientific facts, 75, 78–79; source of, 103
Values Party: New Zealand, 167
Verdi, i (Italian), 168
Verts, les (France), 168

Wallis, John, 50
Water and hydraulic power, 204; dams, 73–74, 171–72
Water pollution, 19–20, 76

Watts, Alan, 88
Ways of My Grandmothers, The (Hungry Wolf), 120–21
Way, The, 100
Wealth, 37, 47
Webb, Walter Prescott, 88–89
Wetlands, 147, 217. *See also* Water pollution
Whales, opposition to hunting, 176
Whitehead, Alfred North, 88, 127–28
White, Lynn, Jr., 66
Whole, relation to parts, 65t., 76
Wholeness and the Implicate Order (Bohm), 77
Wicca, 117–19; critique of, 119–20
Wilderness: conservation or restoration of, 87, 124, 159. *See also* Forests; Nonliving beings; Plants
Wildlife, 134, 176–77. *See also* Animals; Nonhuman beings
Winter, Paul, 99
Winthrop, John, 66
Wirth, Timothy, Senator, 18
Witchcraft. *See* Wicca
Women, 208; and farming, 202–3; of Latin America, 204–5; linked with nature, 59, 184, 190–91, 193–94, 197; of the Second World, 207–9; under capitalism, 197; as workers, 204, 206, 230. *See also* Ecofeminism; Feminism
Women of All Red Nations, 193
Workers: under capitalism, 33, 59, 133; women as, 204, 206, 230
Work ethic, 2
World Bank, 228
World Commission on Environment and Development, 227
World Council of Churches, 122–23
"World machine", 48–49
Worldviews: dominant Western, 88–89, 90–91t.; and environmental paradigms, 85, 88–89, 92f.; indigenous, 80, 120–22, 248–49n.44; "resourcists", 229–30; and science, 41, 51. *See also* Cosmologies; Mechanistic worldview; Organic worldview
Worldwatch Institute, 230
Wren, Christopher, 50

Yugoslavian Green Union, 172

Zen Buddhism, 76
Zimbabwe, ecofeminism in, 203–4